BIOTECHNOLOGY IN AGRICULTURE, INDUSTRY AND MEDICINE

CELLULOSE COMPOSITE FOR ELECTRONIC DEVICES

Biotechnology in Agriculture, Industry and Medicine

Additional books in this series can be found on Nova's website under the Series tab.

Additional e-books in this series can be found on Nova's website under the e-book tab.

BIOTECHNOLOGY IN AGRICULTURE, INDUSTRY AND MEDICINE

CELLULOSE COMPOSITE FOR ELECTRONIC DEVICES

SARUTE UMMARTYOTIN
AND
MOHINI SAIN

nova publishers
New York

Additional color graphics may be available in the e-book version of this book.

Library of Congress Cataloging-in-Publication Data

Names: Ummartyotin, Sarute, editor. | Sain, Mohini, 1956- editor.
Title: Cellulose composites for electronic devices / editors, Sarute Ummartyotin (Department of Chemistry and Materials Technology, Faculty of Science and Technology, Thammasat University, Thailand), and Mohini Sain (Department of Physics, Faculty of Science and Technology, Thammasat University, Thailand, and others).
Description: Hauppauge, New York : Nova Science Publishers, Inc., [2016] | Series: Biotechnology in agriculture, industry and medicine | Includes bibliographical references and index.
Identifiers: LCCN 2015043131 (print) | LCCN 2016000833 (ebook) | ISBN 9781634841733 (hardcover) | ISBN 9781634842082 (Ebook) | ISBN 9781634842082 ()
Subjects: LCSH: Organic semiconductors--Materials. | Light emitting diodes--Materials. | Cellulose fibers. | Composite materials.
Classification: LCC TK7871.99.O74 C45 2016 (print) | LCC TK7871.99.O74 (ebook) | DDC 621.3841/340284--dc23
LC record available at http://lccn.loc.gov/2015043131

Published by Nova Science Publishers, Inc. † New York

CONTENTS

PREFACE

For many years, we have been teaching and researching bio-based materials, bio-based composite and its engineering application at undergraduate and graduate levels, conducting research in a wide range of bio-based materials, such as polymer based composite as well as renewable and sustainable materials. Furthermore, our work also applies to the development of commercial products and systems relating to automotive parts, infrastructure, medicine and electronic devices. This book has been developed from our collective knowledge on bio-based materials for electronic devices.

The concept of bio-based materials for electronic devices is partly an art and partly a science – both are important for the development of effective and efficient electronic devices. Nevertheless, up to the present time, we observe that many commercial electronic devices prepared from conventional techniques are developed largely by empirical or trial-and-error methods. The practice of bio-based materials for electronic devices with little consideration of the relevant science frequently leads to ineffective designs and poor products. In this book, we attempt to strike a healthy balance between art and science, hoping to equip the reader with the necessary knowledge and tools to tackle bio-based materials for electronic devices with ease.

This book is unique in several aspects. Most noticeably is its richness in illustrations, examples, discussion questions, and case studies, which are useful learning tools but are either absent or few in quantity in other bio-based materials books. Another unique aspect is the breath and depth of materials, covering the basic principles and technologies as well as more advanced or current topics such as bio-based materials for electronic devices.

The content of book is divided into six different chapters.

Chapter 1 provides an overview of bio-based materials. It discusses the significance and the feasibility of bio-based material as well as the design of bio-based composite. In addition, we discuss the application of bio-based material for electronic devices.

Chapter 2 is related to the requirement of flexible substrates. Since the development of bio-based materials is the topic of this book, the uses of appropriate substrates are reviewed. The content of this chapter is related to properties of plastic based substrates, particularly the difference between flexible substrates and flat substrates.

Chapter 3 discusses cellulose preparation. The source of cellulose is versatile: it can be prepared from wood based cellulosic material or it can be synthesized by bacterial species. Various chemical and physical treatments have been employed to modify cellulose for use in many applications. Furthermore, the development and properties of cellulose have been declared as being flexible and transparent substrate with the additional feature of touch-screen.

Chapter 4 discusses the operation of organic light emitting diodes, which was considered an optoelectronic device. It operates by the combination of hole and electron transportation from cathode and anode layer. The combination of both hole and electron is called an exciton and it occurs in an emissive region and it subsequently provided the light.

Chapter 5 is related to electronic material for being used as an anode, cathode and emissive layer. Numerous materials have been investigated for use as an electronic circuit in organic light emitting diode. The example was due to metal nano-scale particle, conductive polymer and its chemical and physical modification.

Finally, Chapter 6 discusses the research directions that need to be explored to make bio-based materials more widely used for electronic devices. Although the use of bio-based material as electronic substrate is interesting since it is non-toxic, easy to use and has low cost of processing, the use of bio-based material is still limited in particular areas of research. In order to use bio-based material with higher efficiency, the advantages and limitations of bio-based material needs to be better understood.

ABSTRACT

The creation of bio-based material is partly an art and partly a science. Both aspects are important in the development of effective flexible substrates in organic light emitting diodes. In the past, the development of flexible substrates has been done largely by empirical techniques creating transparent plastic substrates. The application of plastic based materials is still limited. Consequently, bio-based materials have been developed to replace the use of transparent plastic substrates for electronic devices. Interest in green materials has grown exponentially since this development.

This book is unique in several aspects. Most noticeably, is its richness in illustrations, examples, discussion questions, and case studies, which are useful learning tools but are either absent or few in quantity in other books on bio-based materials. Another unique aspect is the breath and depth of material, covering the basic principles and technologies as well as more advanced or current topics such as bio-based materials for electronic devices.

ACKNOWLEDGMENT

Much of the material has been developed from our lectures, conference presentations, research publications and consulting assignments. We thank many of our students, colleagues and friends who have either directly or indirectly contributed to this book. We especially thank Dr. Hathaikarn Manuspiya, The Petroleum and Petrochemical College, Chulalongkorn University, who helped edit the manuscript. Dr. Julasak Juntaro and Dr. Natthaphon Bunnak for sharing their research experience. Most importantly, we thank our families for their support and patience during the writing of this book.

Sarute Ummartyotin
Mohini Sain

About the Authors

Sarute Ummartyotin

Sarute Ummartyotin received a Diploma in engineering and Ph.D. degree from the Ecole Nationale Supérieur de l'Ingenieur de Limoges, France, and the Petroleum and Petrochemical College, the Chulalongkorn University, Thailand, in 2008 and 2012, respectively. His studies are centered on materials science and engineering. Specifically, his research is focused on polymer and ceramic science, materials chemistry, bio-based materials, and bioenergy. He has been employed as a Postdoctoral Fellowship Research Associate at the Centre for Biocomposites and Biomaterials Processing at the Faculty of Forestry, University of Toronto, Canada and is currently working as lecturer for the Department of Chemistry and Materials Technology, Faculty of Science and Technology, Thammasat University, Thailand.

Email: sarute.ummartyotin@gmail.com

Mohini Sain

Dr. Sain specializes in nano-biotechnology, biocomposite and bio-nanocomposites at the Faculty of Forestry, University of Toronto, Canada. He is cross-appointed to the Department of Chemical Engineering & Applied Chemistry, and Mechanical & Industrial Engineering. Dr. Sain's research areas include nano-biocomposite manufacturing and durability; reinforced natural fibre composite light-weighting & manufacturing process, including injection; LFT; thermoforming; extrusion; Resin Transfer Molding (RTM); and Sheet Molded Composite (SMC). His research also extends to bio-refinery, bioplastics and platform biochemical isolation and syntheses by chemical and biological pathways. He is part of a global network in bio-refinery and biomaterials research and commercialization spanning from Brazil to South Africa. His research also contributes new concepts in modeling creep behavior of industrial biomaterials, developing cost effective and commercially viable products and technologies from biomass and other renewable raw materials. He also works on surface science of fiber thin film & green electronic devices, nano-papermaking and related areas.

Email: m.sain@utoronto.ca

Chapter 1

OVERVIEW

Since Kodak's breakthrough three decades ago, organic light-emitting diodes (OLEDs) have attracted considerable interest owing to their promising applications in flat-panel displays [1-6]. Numerous extensive efforts have been developed on this novel technology to replace the use of cathode ray tubes [7-10] and liquid crystal displays [11-16]. Historically, cathode ray tubes are defined as a vacuum tube containing one or more electron guns, and a fluorescent screen used to view images. Cathode ray tubes function to create an image by accelerating and deflecting electron beams onto a screen. As a result, an image is created in the form of an electrical waveform or oscilloscope. Devices that use this mechanism to display images include radar as well as television and computer monitors. The application of cathode ray tubes includes memory devices and light emitting materials and it can be further applied to liquid crystal displays and plasma displays. Up to now, the development of cathode ray tubes has been extensively improved to keep up with the advancements of display technologies. Although the use of cathode ray tubes offers many benefits, their use has many limitations. For example, small amounts of X-ray radiation are emitted when operating cathode ray tubes. Moreover, there are substantial environmental concerns since cadmium and lead are toxic and volatile by-products resulting from the electron beam hitting the fluorescent screen, which contains phosphorus.

To overcome the limitations of cathode ray tubes, liquid crystal displays were developed. Fundamentally, liquid crystal displays are defined as electronic devices that use light to modulate properties of liquid crystal. Generally, it cannot emit light directly but it is able to display arbitrary images and fixed images, which can be displayed or hidden. The application of liquid crystal displays is versatile, including computer monitors, televisions, instrument panels and aircraft cockpit displays. Extensive effort has been made to improve liquid crystal displays past the capabilities and limitations of cathode ray tubes. Benefits of liquid crystal displays are due to its low electrical power consumption and safety with regards to environmental impact. In addition, liquid crystal displays are more energy efficient, very compact and light, generate little heat and maintain geometric integrity. As a result, liquid crystal displays have gained considerable interest in being developed as a light emitting device and it enabled further improvements on temporal performance, color features and resolution. Although the use of liquid crystal displays includes numerous advantages of both the quality and quantity of the image, it still requires attention from the research community to improve upon some aspects of its performance, such as viewing angle, contrast, and brightness.

An alternative to liquid crystal devices and cathode ray tubes are light emitting diodes [17-21], which are two-lead semiconductor light sources or a P-N junction diode. Light is emitted when activated by applying voltage to the leads. Activation causes an electron to fill an electron hole within the device and energy is consequently released in the form of a photon. Light emitting diodes are sometimes called electroluminescent materials and the color of light can be controlled by the energy band gap of the semiconductor [22-25]. The role of light emitting diode gained substantial interest in numerous areas of research such as aviation lighting, automotive headlamps, traffic signals, and camera flashes. Moreover, one advantage to light emitting diodes is its size. The operation area is typically small, approximately 1 mm^2, and it can be integrated onto optical components. The latest developments for this technology has focused on uses in text and video display as well as electronic sensors for communication technologies [26-30]. Although the use of light emitting diodes gained interest due to its efficiency and lifetime of use, it is still limited due to the price of raw materials and processing. In addition, environmental sustainability is an issue since the operation involves lead, a hazardous substance that can have negative environmental effects. For engineering technology to be sustainable, the use of hazardous substances or processes that release toxic and volatile compounds should be avoided. To promote sustainable developments in engineering and technology, Green alternatives should be explored to minimize environmental impact. Moreover, to maintain the quality of electronic devices, the use of light emitting diodes should be considered based on their temperature dependence and voltage sensitivity.

Many developments of light emitting devices are focused on the development of organic light emitting diodes, which use organic materials as the emissive electroluminescent layer. The use of organic based materials in light emitting devices is an excellent Green option and its use should be encouraged. The range of organic based material is versatile due to its molecular weight and chemical structure. Molecular weight varies from small molecules to oligomers and polymers and as a result, the chemical structure is versatile. Variety in the molecular weight and chemical structure of the organic material has allowed for differences in the quality of organic light emitting diodes. In addition, the use of organic materials allows the diode to be flexible. For these reasons, many novel applications for display technologies have arisen especially in the development of portable systems such as mobile phones and portable computers.

Organic light emitting diodes have been developing for 50 years. In the 1960s, the phenomenon of electroluminescence, the emission of light from material in an electric field, was observed in a single crystal of anthracene [31]. Despite the high quantum efficiency obtained with such organic crystals, no application has emerged owing to the high working voltage required as a result of the large crystal thickness and poor electrical contact quality. However, this phenomenon has brought some idea on the activation of emissive materials from applied voltage for light generation. Studies from that time led to a good understanding of the basic physical properties involved in organic electroluminescence such as charge injection, charge transportation, exciton formation and light emission.

A major breakthrough was achieved in 1987 when Tang and Van Slyke, employed by Kodak, reported efficient and low-voltage organic light emitting diode from P-N junction devices using thin films of vapor deposited organic materials [32]. Further development of the technology ensued in the 1990s with the discovery of electroluminescence from polymers at the University of Cambridge. The key advantages of organic light emitting diodes for display technology are due to the self-emitting properties, high luminous efficiency, full color capability, wide viewing angle, high contrast, low power consumption, low weight, potentially large area color displays and flexibility. Several approaches have been described to build full-color displays, such as side-by-side patterning of discrete red (R), green (G) and blue (B) sub-pixels, RGB tunable pixels, filtering of a white organic light emitting diode, down-conversion of blue emitting organic light emitting diode, and filtering of broad-band-emitting organic light emitting diode [33-35]. The impressive performance of organic light emitting diodes is not only due to organic material making up the emissive layer but the fabrication technique and electronic substrate also contributes to the performance of the diode.

In terms of sustainable technology, organic light emitting diodes are exceptional since lead is not used as emissive material. Moreover, the flexibility of organic and polymeric-based emissive layer make organic light emitting diodes a special interest for manufacturers of flexible and portable display screens. Consequently, many types of transparent polymeric-based substrates have been investigated as potential electronic substrates to replace the use of glass or mirrors as an electronic substrate. Transparent polymeric substrates are more desirable due to flexibility of the material as well as the cost and processing of the material.

Due to the growth of flexible display technology, more and more development is done to facilitate a diversity of users. Since organic light emitting diodes require a polymeric-based substrate, interest in this product has grown exponentially. Fundamentally, flexibility in display technology is enabled by a non-glass backplane of electronic devices [36-39]. Materials that are used to fabricate the electronic device should be bendable enough to allow folding and rolling. Flexible displays are sometimes called rollable displays. They can gain bendability and flexibility during their period of use. In recent years, the literature on flexible displays has widely expanded, including books on flexible flat panel displays and on flexible displays. The prospects for flexible displays are promising, although application still depends on technical and manufacturing developments. Electrophoretic displays such as electronic papers using plastic substrates, which have a relatively simple structure, are just beginning to be produced in quantities approaching high volume. Flexible display technology that was intended to flex or roll during use may reach the market in several years, pending further developments in backplane and fabrication processes. The near-term revenue dynamic and portable display will lead to the development of larger and more sophisticated displays with flexibility and roll-ability.

In order to reach the full potential of display technology, polymeric substrates are very promising materials for flexible displays with many advantages such as transparency, low weight, flexibility, and robustness. They are a good alternative compared to glass substrates that have been widely used for flat panel displays such as cathode ray tubes, liquid crystal displays, and plasma discharge panel technology. Glass is difficult to use in flexible displays since it is inherently rigid. As an alternative and to facilitate users, polymeric-based substrates exhibit mechanical properties that vary from strong rigidity, such as in engineering plastics, to softness, such as in rubber or commodity plastic. The use of polymeric-based substrate gained lots of interest due to the use of inexpensive materials and the feasibility to be scaled for industrial commercialization. Mass production of display technology was achieved on roll-to-roll (RTR) process. The flexibility of polymeric-based substrates proved to be beneficial compared to glass based substrates [40-42]. Since polymeric-based substrates are flexible, they can be used in many electronic devices and its use in electronic devices was evaluated based on the feasibility of a flexible electronic device. The other electronic device is due to flexible batteries and transistors as well as capacitive material [43-45].

Due to the achievement of polymeric-based substrate, many transparent polymers were investigated as potential flexible substrates. The use of polymeric-based materials as flexible substrates is advantageous not only in terms of flexibility and facilitating between users but it is beneficial due to cost effectiveness, ease of fabrication, and it can be developed for industrial commercialization. Although the use of polymeric-based substrate for flexible display offers many benefits, it is still limited in terms of device operation. The heat generated by electronic devices is considered an important factor in determining the lifetime and performance of the device. Heat generated from an electronic circuit when the device is operated can be lead to device failure. The quality of the device and its lifetime are therefore detrimental. This concern has pushed scientists and researchers to further develop flexible substrates.

Up to the present time, numerous efforts have paid explicit attention to polymeric-based composite materials. Due to population growth and high production and consumption demand of petrochemical based polymeric materials, the need for petrochemical based plastic was considered and a shortage is expected in the near future [46-48]. In recent years, concern over the shortage of petrochemical materials and the ever growing issue of global waste, along with rising oil prices and the pending exhaustion of landfills, has created a demand for research and development of eco-friendly materials. Numerous efforts have investigated the feasibility of bio-based materials in order to replace the use of petrochemical based products. The advantage of bio-based materials was strongly associated with non-toxic properties and environmentally friendly purposes. From the fundamental point of view, bio-based material is defined as a substance derived from a living organism. Bio-based material is versatile, depending on its life source such as wood and leather, and its properties can be further extended. One successful project using bio-based materials focused on engineering wood and its products, which were involved in oriented strand board and particle board [49-52]. From the fundamental point of view, the use of wood engineering should be strongly encouraged and developed since it is a naturally occurring, abundant bio-based material. Wood engineering has therefore chemically and physically modified wood for use as cellulosic based materials. To isolate cellulose and its impurities such as lignin, pectin and any derivatives, it was prepared and modified for most possible applications [53-56]. For example, cellulose exhibits many advantages such as strength, stiffness, chemical and thermal resistance and it gained interest for the development of engineering applications. The use of lignin is of interest for use as an adhesive material, however the contact angle between interface regions is acute if lignin is employed as adhesive material. Up to the present time, ample research efforts have focused on many bio-based materials. It can be covered to chitin-chitosan, polylactic acid, and polybutylene succinate as well as starch and soy protein.

For electronic devices, flexible display substrate is capable because of bio-based materials. One alternative approach uses cellulosic based material to obtain a flexible substrate [57-60]. The use of cellulose is satisfying in terms of thermal resistance. Low coefficient of thermal expansion on cellulose is excellent and it is suitable for further development as an electronic substrate. Moreover, one challenge to the use of cellulose is its application at the nano-scale network since cellulose is scaled down from a bulk particle to a nano-scale network. The size of cellulose is much smaller than a wavelength of light and, therefore, light passes through cellulose without any diffraction. In order to take advantage of the excellent properties of cellulose including the low coefficient of thermal expansion, its chemical resistance, optical properties, and mechanical properties, development is required for use as a flexible display substrate. As mentioned in previous literature, the co-efficient of thermal expansion of cellulose was estimated to be 0.1 ppm/K. Excellent mechanical properties can be explained by Young's modulus of its single fibril of 114 GPa. It also has attractive features of high degree of crystallinity of 89, a high degree of polymerization of 14,400 and a high specific area of 37 m^2/g. Furthermore, an excellent feature of chemical resistance, high dimensional stability as well as stiffness was one of the criterions in order to use cellulose in an electronic display substrate.

From a structural point of view, cellulose can be effectively derived either from a top-down approach, in which biomass was subjected to high shear forces in order to create a smaller size of cellulose in suspension, or from a bottom-up approach, utilizing the biosynthesis of cellulose by bacteria, in which the most effective bacterial species was Acetobacter Xylinam [61-63]. The advantage of bacterial cellulose was related to the purity of the product. Cellulose prepared from bacteria is free from wax, lignin, pectin and hemicelluloses, which is commonly presented in cellulose derived from plants. Moreover, cellulose prepared from bacteria can control the repeating unit and the molecular weight of the fermentation process. However, from the viewpoint of industrial commercialization, the cost of cellulose prepared from bacteria is relatively high. The use of bacterial cellulose-based material for display technology substrate is therefore limited in terms of mass production. From the viewpoint of industrial commercialization, cellulose derived from plants gained interest due to cost effectiveness and availability of raw material. The cellulose derived from plants is considered waste from agricultural production and developments to attain value-added products are desirable. In industry, cellulose is prepared from plants. Chemical and physical modification of the properties of cellulose is done to remove any impurities. Variation in size and shape of materials is therefore modulated. Chemical treatments change the properties of the cellulose surface while variations in size can be controlled by physical modification. However, the use of cellulose is still limited and requires further modification and improvements.

One of limitations of the use of cellulose-based material for electronic devices regards the water and gas absorption ability of cellulose. Cellulose derived from plants is involved in extraction and chemical modification processes. The use of water was preferred because of the water-based suspension of cellulose. Controlling the dispersion and distribution of cellulose particles in water-based solutions is an important consideration. As a result, it can be noted that the cellulose preparation process is completed in water and its absorbability is due to its hydroxyl group. The device quality and lifetime is therefore dependent on whether water is generated by the device, which is detrimental. Any concern about the use of cellulose based flexible display substrate is related to the surface roughness of cellulose. Compared to glass, which is limited due to brittleness, and its processing is limited in the case of roll to roll, the smoothness of glass sheets are excellent. Glass has therefore gained interest as a substrate for conductive thin film formation. Epitaxial film should preferably be grown on a smooth surface, which determines device quality and lifespan. Because of these reasons, more and more research has exclusively focused on minimizing the disadvantages of cellulose for use as a substrate for flexible display technology.

To overcome the limitations of cellulose, it is physically modified into a composite material. From the structural point of view, composite based material is defined as materials made from two or more constituent materials with significantly different physical and chemical properties. The combination of two or more different materials offers significant change in novel structure rendering properties. In the case of cellulose, the concern over water sorption ability and surface roughness can be tackled by forming composite materials. Up to the present time, many types of bio-based materials have been investigated for feasibility of application. The challenge of cellulose and its modification is very far away from the forest. Many developments in medicine, pharmaceutical technology, food engineering, and infrastructure have found applications in Green technology.

Chapter 2

PREREQUISITE REQUIREMENT ON SUBSTRATE

The development of organic light emitting diodes gained interest due to their ease of use and capacity to facilitate between users. More and more efforts have contributed to the continual developed of this achievement. One challenge to constructing a flexible display is the development of a flexible substrate. It is important to note that the flexible substrate should be well designed for display technology. However, in order to understand the criteria of flexible display substrate, attention should be given to polymeric-based substrate. Questions involve the design, or how to design and any advantages, or how to overcome any issues. From the fundamental point of view, the consideration of flexible display substrate involves polymeric-based materials. It is very important to note that thermal and chemical resistance and water absorption to the polymeric substrate is an important issue that we need to consider [64].

To support the use of polymeric-based substrate for flexible display technology, it can capture much interest due to its processing. In the past, the traditional light-emitting device used as substrate was glass sheet, which resulted in many limitations due to non-flexibility, and difficulty of processing. Also, the processing was still slow due to the non-continuous process. Due to glass sheet the fabrication process was still limited and only one approach on the film formation process relied on batch reaction [65, 66]. The film growth on the substrate should be considered a vapor deposition process [67-70]. It can be involved in both physical and chemical vapor deposition. The example of physical vapor deposition was due to thermal evaporation [71], E-beam evaporation [72], etc. On the other hand, the example of chemical vapor deposition was due to thermal chemical vapor deposition, plasma enhanced chemical vapor deposition, etc. [73, 74]. Although the use of both chemical and physical vapor deposition gained much interest due to film quality and efficiency, the use of this technique was still limited in processing conditions. From the fundamental of view, the use of vapor deposition required high pressure and a controllable process for being a vacuum system. It resulted in high cost of processing, and safety and economical concerns for being a mass production process for industrial commercialization. Figure 1 and Figure 2 exhibit the process of chemical and physical vapor deposition processes.

On the other hand, the use of the vapor deposition process employed high temperature for activating the film source. As a result, it was therefore limited on particular substrate that can withstand high temperature. The example of substrate was due to metallic and ceramic substrate. It can be implied that vapor deposition commonly utilized metallic and ceramic substrate and its application was involved in electronic circuits when high temperature resistance was required. As a result, in order to employ polymeric-based substrate, it was expected to be flexible because the nature of polymer was still limited. Up to the present time, many developments on the vapor deposition process were focused on the reduction of temperature for processing. The use of plasma physics or discharge on dielectric barrier was being investigated and it resulted in lower temperature of processing. The use of high pressure and non-continuous process were still issues that made device efficiency limited and industry commercialization was not realistic. Figure 3 exhibits the schematic diagram of the plasma enhanced chemical vapor deposition process.

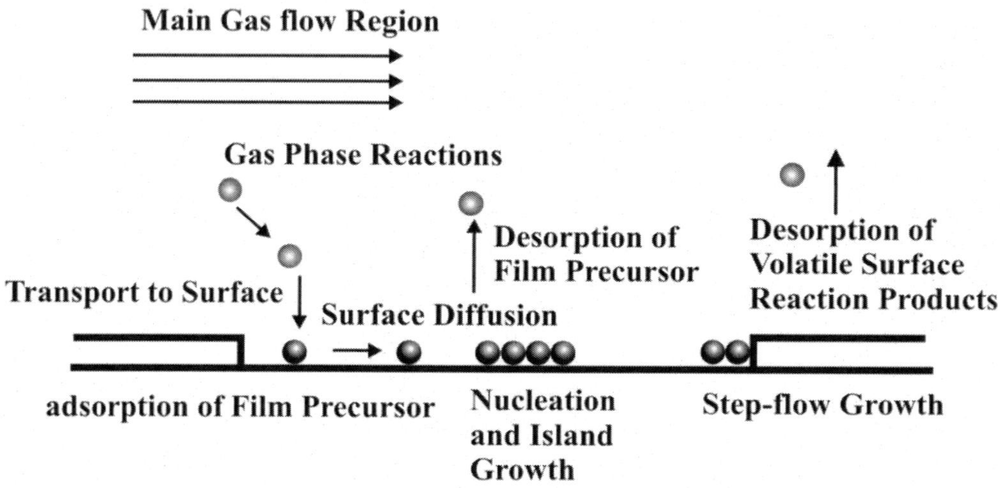

Figure 1. Chemical vapor deposition.

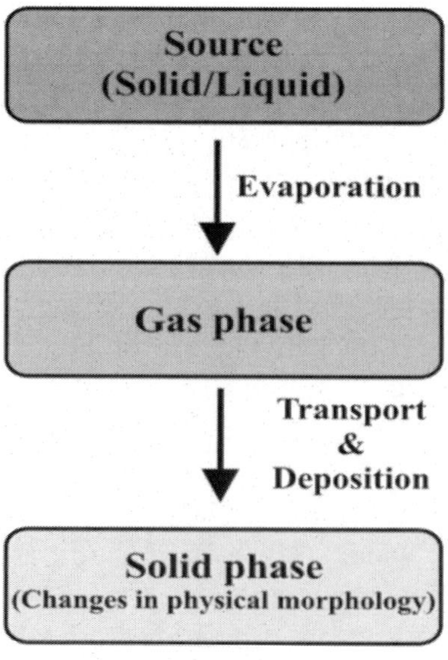

Figure 2. Chemical vapor depositions.

In this chapter, the main content involved the development of flexible display substrate. Before understanding flexible display substrate, a prerequisite requirement on flexible substrate should be declared.

2.1. Potential Polymer Candidate

Due to the push to develop flexible display substrate, many investigations were focused on the development on polymeric-based substrate. Several polymers have been investigated for this feasibility and it was important to note the preliminary criterion was based on transparency and thermal resistance [75]. Thermal characteristics and their affects on the flexibility of polymer should be understood. It can be noted that thermal behavior of polymeric-based materials can be categorized into three different regions. At low temperature, the polymer became rigid and hard. If elevated temperature was applied, the polymeric chain was motivated and it became flexible. It can be said that the status of polymer became rubbery and soft. Later on, if higher temperature was applied, the temperature was therefore melt. At this stage, the occurrence involved an exothermic region. However, if the temperature was then cooled to room temperature, the polymer chain could be re-crystallized. The crystal became re-oriented. Figure 4 exhibits the thermal characteristics of polymeric-based materials.

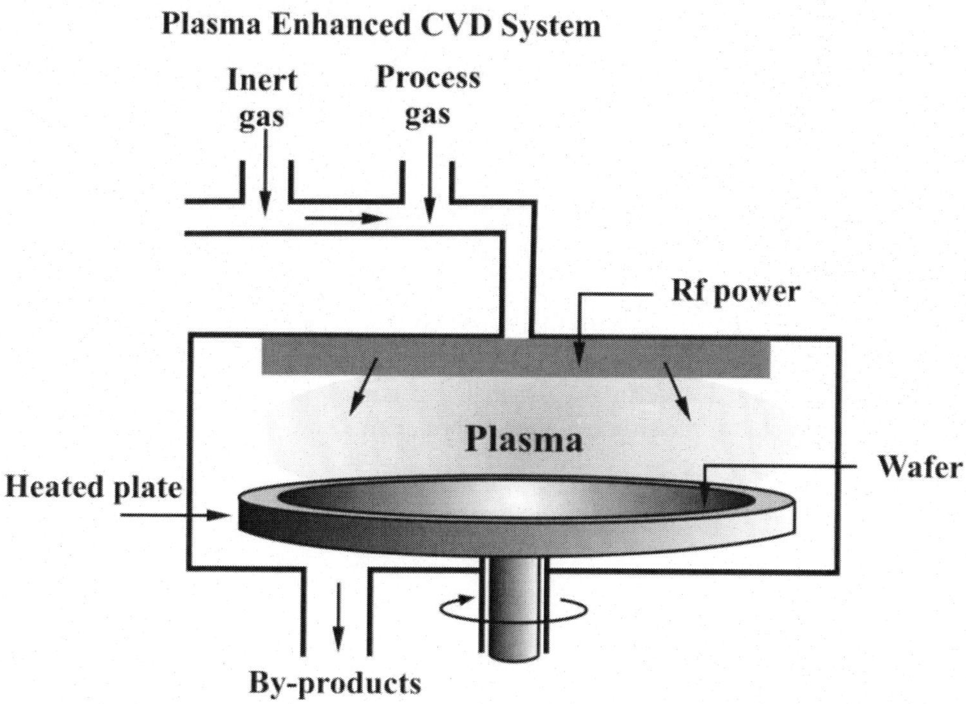

Figure 3. Plasma enhanced chemical vapor deposition process.

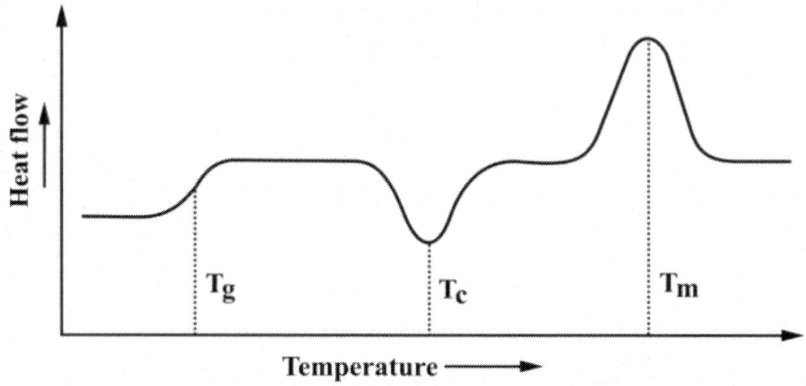

Figure 4. Exhibits the thermal characteristic of polymeric-based materials.

Due to the thermal characteristics of polymeric-based materials, thermal sensitivity can be implied. One of the challenges of the design of polymeric-based substrate for display technology concerned the co-efficient of thermal expansion of substrate, which was commonly employed for high temperature resistance. From the fundamental point of view, it can be noted that polymeric-based substrate can be categorized into two different types [76]. One was focused on the thermoplastic, while the other one was focused on thermoset. Thermoplastic was considered a typical polymer. It became pliable or moldable above the glass transition temperature and it can be solidified upon cooling. In general, thermoplastic has high molecular weight. The polymer chain connects through intermolecular forces, which weaken rapidly with increased temperature and yield a viscous liquid. The structure of thermoplastic commonly referred to linear chain. However, on the other hand, thermoset was considered a petrochemical material that was irreversibly cured. The cure can be induced by heat or UV through a chemical reaction or irradiation [77-79], respectively. It can be noted that the polymerization for thermoset typically provided the small molecule of by-product and it can be implied that it was irreversible polymer. The use of thermoset was usually liquid or malleable before curing and designed to be molded into its final shape and it was commonly employed as adhesive. The structure of thermoset commonly referred to branch and network polymer. It therefore offered high thermal resistance. For flexible display substrate, the role of thermoset was satisfactory in terms of thermal resistance, but its use was still limited in flexibility. Figure 5 exhibits the structure of polymer chain.

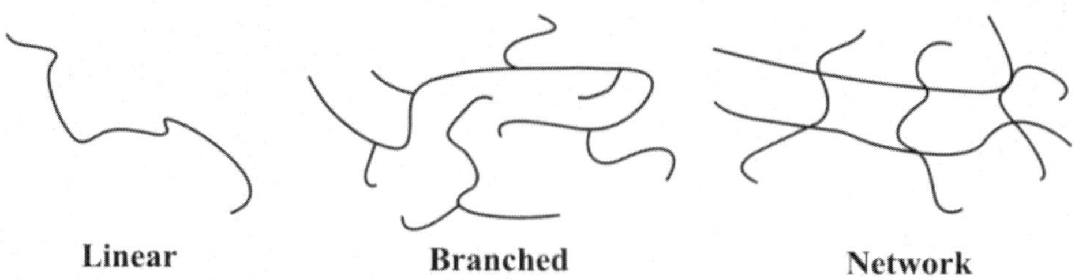

Figure 5. Exhibits the structure of polymer chain.

Table 1. Glass transition temperature and application of polymer

Polymer	Glass transition temperature (°C)	Application
1. Polyethylene terephthalate (PET)	70	Transparent bottle
2. Polyethylene naphthalate (PEN)	140	Photonic device, substrate
3. Polycarbonate (PC)	140	Transparent sheet, Food packaging, Wire
4. Polyethylene sulphone (PES)	180	Dielectric and capacitive materials
5. Polycyclic olefin (PCO)	200	Packaging, health care, optical device
6. Polyarylate (PAR)	200	Flame retardant
7. Polyethetherketone (PEEK)	340	Automotive part
8. Polyimide (PI)	200	Electronic device

Up to the present time, many types of polymeric-based materials have been investigated on their feasibility to be employed as flexible substrate of display technology. The example of substrate was due to polyethylene terephthalate (PET), polyethylene naphthalate (PEN), polycarbonate (PC), polyethylene sulphone (PES), polycyclic olefin (PCO), polyarylate (PAR), polyetheretherketone (PEEK) and polyimide (PI), respectively. The glass transition temperature and application of polymer are exhibited in Table 1 and the molecular structure of an example of polymer is exhibited in Figure 6, respectively.

Up to the present time, a transparent polymer was chosen and investigated on the feasibility to be employed as flexible display technology. Although transparent polymers were investigated on their feasibility to be flexible substrate, it was stilled limited due to low thermal resistance [80, 81]. Excellent substrate for flexible display technology should be preferably considered on the co-efficient of thermal expansion (CTE). One of the highest concerns regarding the use of flexible polymer should be control. Heat was generated and it affected the quality of the device and its lifetime [82, 83]. Moreover, a novel approach on display technology not only focused on flexibility but also on the ability of the touchscreen, being pushed for academic research and industrial commercialization. It can be noted that the development of touchscreen can be facilitated in between users.

From a fundamental point of view, the touchscreen was considered an input device, normally layered on top of an electronic visual display of an information processing system. A user can give input or control the information processing system through simple or multi-touch gestures by touching the screen with a special stylus/pen and/or one or more fingers [84-86]. This concept of touchscreen was interesting due to facilitating for user, fast input process and response. The feature of touchscreen is versatile in many applications such as mobile phones, and computers as well as books. In the past, the touchscreen feature was considered an additional item for facilitating and useful in between users. To gain the touchscreen feature, film formation was grown on substrate and as a result, it was then employed. It seemed that the ability of a touchscreen was developed after flexible substrate, similar to the finishing process for plastic.

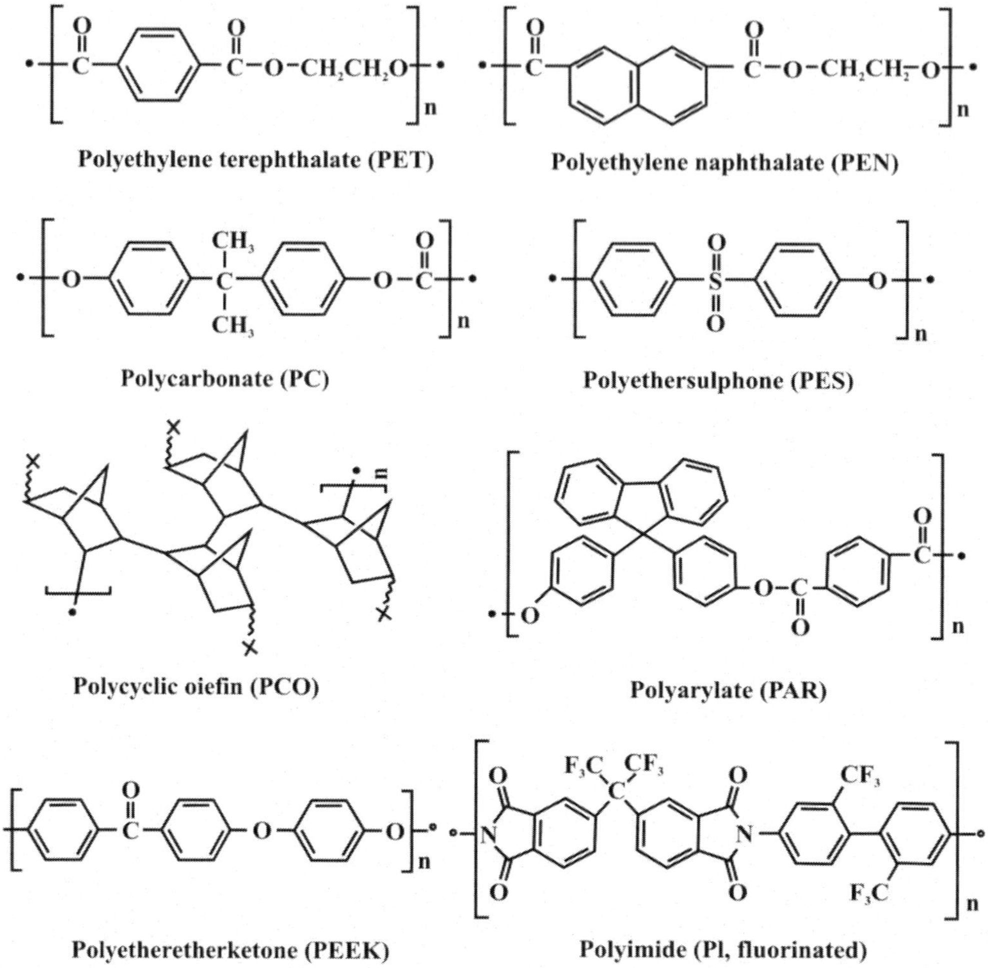

Figure 6. Molecular structure of transparent polymer for flexible display technology substrate.

In recent years, touchscreen devices have become important in commercial electronic applications due to their simple and intuitive use. Touchscreen is an electronic visual display that users can directly control by touching the screen rather than using a mouse, touchpad, or other controller devices. Touchscreen technology has been used in computers since the 1960s. In fact, there are several types of touchscreen technology, which are used in familiar electronic devices such as automatic teller machines (ATM), inventory control, and even cashier computers in convenience stores. Using touchscreen technology, using fingers to control these instruments directly, was more convenient than using a keyboard or mouse. About ten years ago, a "personal digital assistant" (PDA) became the most outstanding electronic instrument, which used touchscreen technology. It was controlled by its stylus. However, this technology was still not interesting enough. Recently, touchscreen technology become highly interesting due to the Apple Inc.'s innovation, "The iPhone." "iPhone phenomena" stimulated the touchscreen technology in electronic applications. Many companies tried and developed their own touchscreen systems electronic devices. Currently, most common electronic devices such as smartphones, tablets or even personal computers, all contain touchscreen technology. The unlimited demand and popularity of these electronic devices are driving forces of touchscreen technology development in the future. Normally, there are three components used in touchscreen technology: First, "the touch sensor", which is a panel with a touch responsive surface. Generally, a touch sensor has an electrical current running through it, and touching the screen causes a voltage change. The voltage change signals the location of the touch. Second, "the controller" is the hardware that converts the voltage changes on the sensor into signals, which the computer or other device can receive. Third, "software", which tells the electronic devices what's happening on the sensor and takes the information coming from the controller. Then, it will allow the electronic devices to respond accordingly. Walker reported that there were several types of touchscreen systems, but only two types are commonly used in electronic devices. Those are capacitive, resistive, and piezoelectric touchscreens.

Capacitive touchscreen or projected capacitive was the most common touchscreen used in many electronic devices [87]. However, this type of touchscreen technology was little known until Apple Inc. used it in the first iPhone in 2007. This capacitive touchscreen had many outstanding properties such as high transparency, abrasion resistance and durability due to its conductive layer covered glass at the surface of the screen. The fundamental theory of this type of screen was the electrical voltage disturbing or changing of capacitance of the instrument due to the capacitance of humans. Figure 6 shows the fundamental theory of capacitive touchscreen works. All four corners of the screen generate electrical voltage to create an electromagnetic field. The electromagnetic field intensity was regular in all areas. When human fingers touch the screen, the electrical voltage decreases. The detector can locate where the point is, then the electrical signal is transformed and sent to the processing unit.

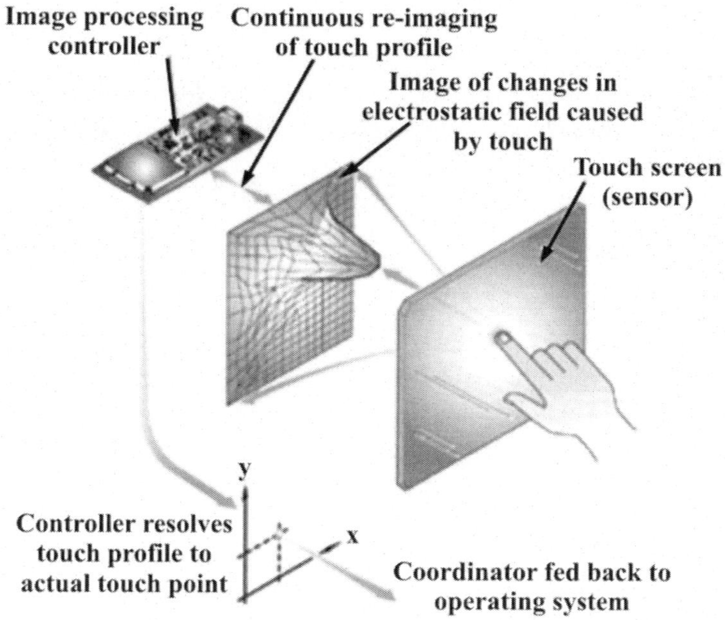

Figure 7. The structure of capacitive touchscreen.

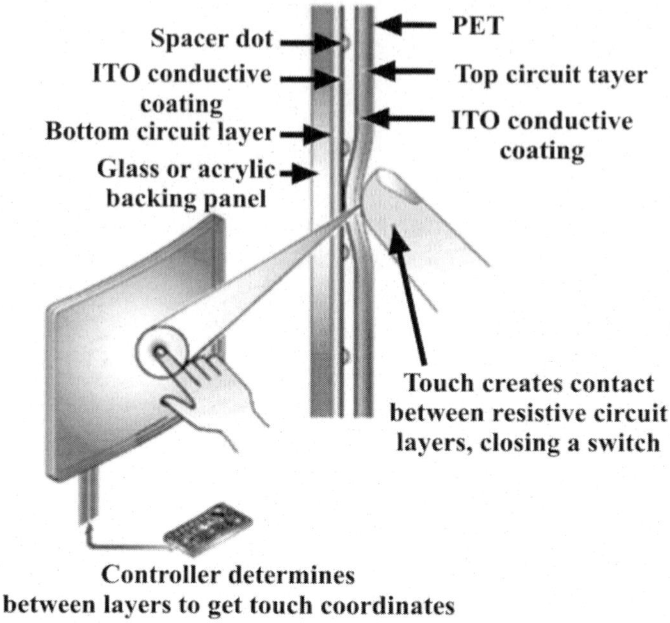

Figure 8. The structure of resistive touchscreen.

The Resistive touchscreen or analog resistive has the 2nd market share among touchscreen systems. It was made by Elo TouchSystems in 1971. The original resistive technology was used in an opaque pen digitizer. A resistive touchscreen is simply a mechanical switch mechanism used to locate a touch. The construction of a typical resistive touchscreen is shown in Figure 7. A lower layer of glass substrate and an upper layer of flexible film (generally PET) are separated by very small, transparent, insulating spacer dots. One side of both glass and PET are coated with a transparent conductive layer of indium tin oxide (ITO). The coated sides are facing each other. The electrical voltage is applied to one or both of the sheets. When a finger presses on the PET film, the two conductive surfaces make an electrical contact. As the resistance between the two conductive layers becomes lower, the detector will locate the point and then send the signal to the processing unit, the same as the capacitive touchscreen system.

The Piezoelectric or force-sensing touchscreen is the touchscreen system that has the piezoelectric fundamentals [88, 89]. From the Greek language, piezo- means stress and electric means electricity. When these two words are combined, the new word "piezoelectric" means electricity generated and applied by stress. Similar to the capacitive touchscreen system, the electrical signal generated by stress may disturb the electromagnetic field on the screen. The detector can locate the point and send the signal to the processing unit. Thus, this touchscreen system can be used for multiple simultaneous touches as a capacitive touchscreen. Moreover, the piezoelectric touchscreen can give a precision location and can be used with any object as a resistive screen due to the stress that must be given. In 2006, patented force-sensing touch technology was developed. This technology contains the advantages of both capacitive and resistive touchscreens. Figure 9 shows the diagrams of the transparent force-sensing layer incorporated in a digital resistive multi-touch screen construction. The fundamental is the same as resistive touchscreen.

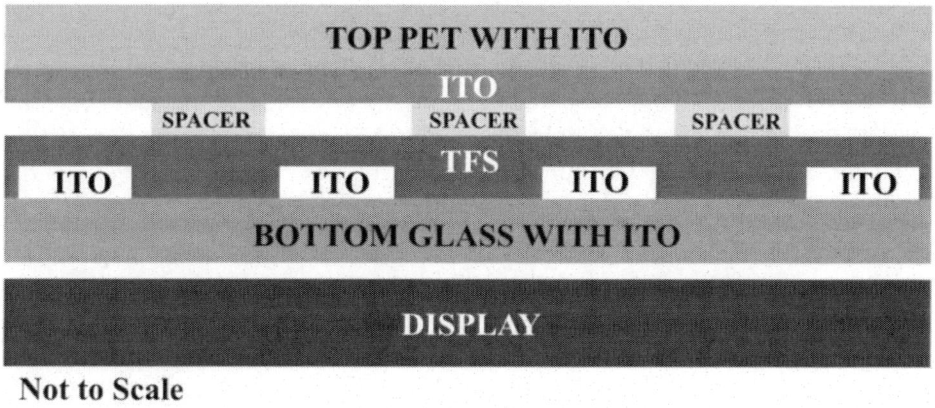

Figure 9. Transparent force sensing layer incorporated in a digital resistive multi-touch screen construction.

In advance, the superior advantage of piezoelectric touchscreen over capacitive and resistive touchscreen is that it produces many levels of pressure, which can control the intensity of the signal with the pressure-sensor. Recently, Apple Inc. revealed a patent of the pressure-sensitive touchscreen. This idea allows people to wonder and predict what the next product from Apple Inc. will be. The advantage of the touchscreen is very convenient to the user. In order to extend the lifetime of touchscreens, materials should preferably be designed to be flexible and transparent.

2.2. Properties of Flexible Substrate for Display Technology

Up to the present time, many transparent polymers have been investigated as to their feasibility as flexible display substrate. Transparency and flexibility are considered the most stringent conditions for flexible display devices.

Moreover, in order to replace the use of glass sheets, polymeric-based substrate was required to mimic the properties of glass, including its clarity, dimensional stability, thermal and chemical resistance, barrier properties against moisture and volatile gas, solvent resistance as well as smoothness of surface, and the most important was ease of processing [90, 91]. In order to be an excellent substrate, the minimum properties of polymeric-based substrate should be restricted and considered as follows.

2.2.1. Clarity

One of the most important characteristics of substrate for optical devices focused on clarity. It was important to note that the first criteria of substrate should involve the ability to transmit the light from the emissive layer. Since the substrate was opaque, the light transmission was limited and it resulted in non-efficiency of the optical device. Figure 10 exhibits the organic light emitting diode structure. The operation can occur from the applied voltage on both sizes of the electrode. The electron and hole were therefore moved from the anode and cathode, respectively. The combination of both electron and hole in the emissive layer zone allowed light to occur. Transmission of light then passed through the flexible substrate. Therefore, it was remarkable that the transparency of substrate could be controlled. It was suggested that the amount of light that should be passed be a minimum of 90% over the wavelength region of 400-1000 nm. This region of wavelength was considered due to visible light region.

2.2.2. Thermal Resistance

Due to the application of the optical device, heat generation was an important concern. From the fundamental point of view, it was remarkable to note that light processing involved an electronic circuit and heat generation consequently occurred. Due to this concern, the process of heat transfer should be understood in order to design flexible substrate.

Heat transfer involved the transfer of heat due to the temperature difference and it can be categorized into three different mechanisms. One focused on conduction. It referred to heat transfer that occurred across a stationary solid or fluid in which a temperature gradient existed. The other one was focused on convection. It therefore referred to the heat transfer that occurred across a moving fluid in which a temperature gradient existed. The last one was due to radiation. It involved the heat transfer between two different surfaces at different temperatures separated by a medium transparent to the electromagnetic waves emitted by the surface. An example of heat transfer is exhibited in Figure 11.

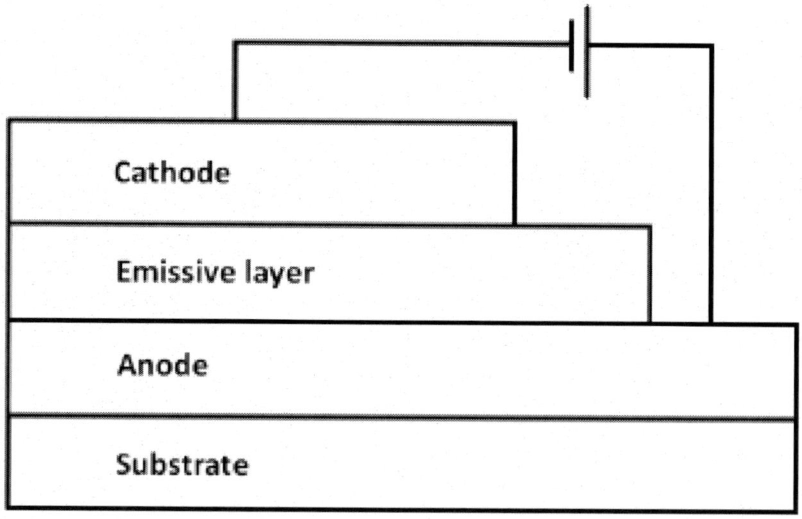

Figure 10. Organic light emitting diode structure.

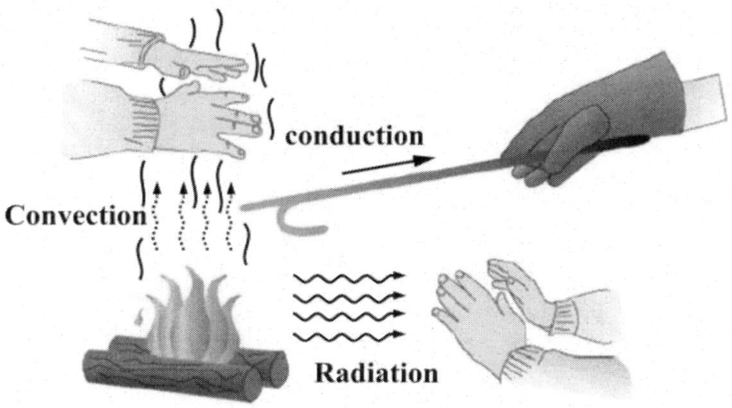

Figure 11. Mechanism of heat transfer.

Due to the application related to the optical device, the design of the circuit and substrate and the heat transfer process should be considered. In flexible display substrate, polymeric-based substrate was employed and it resulted in concern about thermal resistance. The correlation between thermal resistance and flexibility should be considered. Polymeric-based substrate should satisfy the low co-efficient of thermal expansion. Moreover, not only for the substrate, the matching co-efficient of thermal expansion in between conductive layer and substrate should be developed. Because of the processing step, polymeric-based substrate should be able to withstand very high temperature. It commonly involves film formation and device annealing from the high temperature of processing.

One concern about the use of plastic as substrate for flexible display substrate involves the match between the co-efficient of thermal expansion between substrate and any conductive layer. To use metal as a conductive layer, it was remarkable that although metal provided significant conductivity, high thermal expansion should be considered. The use of a ceramic-based layer may be used to control the pathway of the electron and hole; crack phenomena should be avoided. Up to the present time, the correlation between thermal expansion, conductivity of an organic layer and feasibility of processing should be evaluated for the display technology plan to be up-to-scale for industrial commercialization.

2.2.3. Surface Properties

Most optical devices focused on film formation. The structure of this device was designed as thin film. Many developments have been investigated regarding their feasibility of use as film formation. Up to the present time, other approaches on optical device were being considered as to their feasibility in industrial commercialization. Film formation gained much interest due to ease of processing, facilitating users and for economical reasons. It was therefore remarkable to note that if film formation was employed, the development on substrate should be considered. The quality of light and efficiency of device lifetime has become much improved, as attention was paid to the quality of substrate.

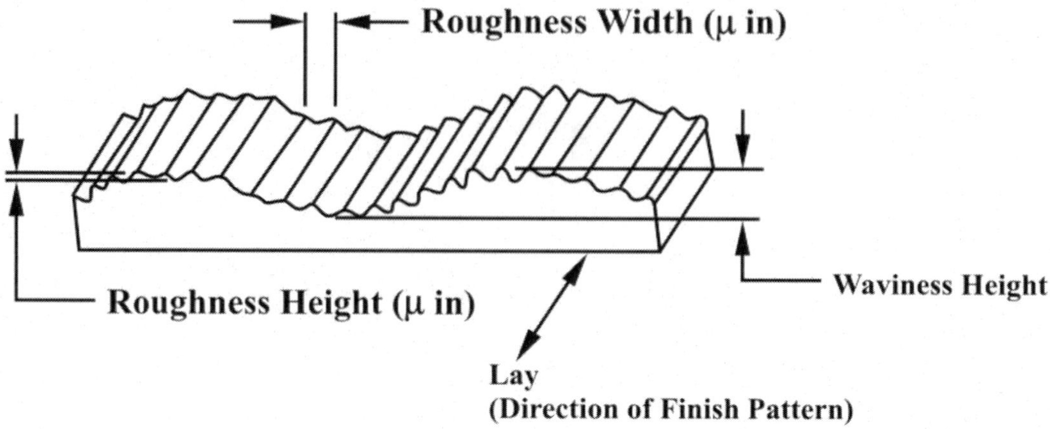

Figure 12. Roughness of substrate.

In order to gain higher efficiency of the optical device, thin film formation was therefore considered one of the most effective routes. The time for mobility of negative change and the positive charge from electrodes should be matched on the emissive layer region. It was remarkable to note that the velocity of charges can be controlled by insertion of any sub-layer commonly fabricated in the nano-scale region in between. To gain effectiveness, the compatibility between the conductive layer and surface of substrate should be optimized.

Compared to conventional display technology, flat glass sheets were employed as substrate. The roughness of the glass sheet was estimated to be 4 angstrom and it resulted on the high quality of film formation, quality of light as well as efficiency. Figure 12 exhibits the roughness of substrate.

Up to the present time, display technology focused on the development of the flexible feature. Polymeric-based substrate was then realized. It was therefore important to note that to gain the high quality of light and efficiency, the roughness of polymeric-based substrate should be considered, similar to the use of flat glass sheet [92, 93]. However, the difficulty concerned the bendable and rollable features during the fabrication process and use. Therefore, surface quality such as roughness and cleanliness were essential to ensure the integrity of subsequent layers. Moreover, the ability of water absorption was definitely different between the flat glass sheet and the polymeric-based substrate,. From the fundamental point of view, polymeric-based substrate exhibited the superiority of water absorption compared to flat glass sheet. As a result, the use of polymeric-based substrate should be carefully considered for ease of defect formation. The defects from surface properties, such as pinholes on the conductive layer, can form a dark spot in the device and it can lead to cracks when the display is bent.

2.2.4. Chemical Resistance

The change from flat glass sheets to polymeric-based substrate not only changed the flexibility but the fabrication process was also altered. In the past, flat glass sheet as substrate for optical devices focused on vacuum technology in order to deposit film formation. The example of vacuum technology for film formation involved physical and chemical vapor deposition, respectively. The advantage of vacuum technology gained many advantages such as film quality, less defects, as well as the efficiency of the device. Moreover, the use of vacuum technology was also beneficial due to the solvent-less process. Removal of solvent was one of the issues that needed to be tackled for film quality [94, 95].

However, the use of vacuum technology for optical devices was still limited if mass production was required. In order to gain higher efficiency for mass production, rapid processing and reliability should be considered. From the viewpoint of industrial commercialization, the roll-to-roll process was commonly employed and it involved a conductive solution. Conductive material was prepared for the solution; the effects of the solution must be considered. Therefore, removal of the solvent after the film deposition process should be considered. Up to the present time, the example of solution based processing involved spin coating, dip coating, film casting and it can be up-scale to roll-to-roll process.

Therefore, to gain the efficiency of the flexible display substrate, polymeric-based substrate should preferably be resistant to any chemical reagents and any hazardous products. The degradation of polymeric-based substrate should be prevented.

2.2.5. Mechanical Properties

Mechanical properties for flexible display technology are one of the hot issues that we need to discuss. Compared to traditional optical device, which used flat glass sheet as substrate; it was remarkable to note that the limitation was due to brittleness and it resulted in fragile device lifetime [96-99]. Up to the present time, the alteration to flexible display technology captured much interest, and mechanical properties were therefore discussed. It was important to note that since we considered the use of polymeric-based substrate and mechanical properties of polymer such as tensile strength, Young's modulus and elongation at break should be considered. It can be noted that polymeric-based substrate exponentially gained interest due to the continuous process. The use of flat glass sheet was limited due to batch formation and concern over the use of a vacuum was dangerous if mass production was then employed.

When the devices are bent, the mechanical discrepancy at the interface between the organic and inorganic materials generates mechanical failure in devices. For those devices, inorganic thin films may be the source of the failure due to their brittle properties. For this criterion, all-organic-based devices were the best prospects for rollable displays. However, certain devices require inorganic films and metal nano-scale particles to be fabricated. Polymeric-based composite between metal nano-particle and inorganic film was then employed for optimization between conductivity and film formation. As a result, it is important to understand and improve the mechanical properties above the limitation of hybrid materials. The failure of devices depends on many things, for example, the arrangement, thickness, and properties of conductive layers. Moreover, the processing ability and adhesion at the interface between the different layers under thermal cycling and environmental testing, wetting characteristics, and the ability to withstand flex testing were critical to determining the robustness of the optical device. Flexible display technology has the advantage of mass production through a continuous process. Therefore, it is essential to understand the mechanical properties required of flexible substrates.

2.3. Barrier Properties

In order to enhance the efficiency of device lifetime and quality, the design of substrate should be free from any environmental effects. The difference between flat glass substrate and polymeric-based substrate considered many aspects, such as the change from brittleness to flexibility, significant enhancement on the performance of processing ability, chemical and thermal resistance, etc.

In the past, the use of flat glass sheet was satisfactory in terms of water absorption ability and gas permeation ability. These excellent properties have been realized as outstanding properties of flat glass sheet up to the present time. Although the use of polymeric-based substrate gained much interest due to ease of processing for mass production in industrial commercialization, the limitations of polymeric-based substrate for flexible display technology such as water absorption ability and gas permeation should be discussed and improved To achieve this concept, the idea of super thin glass was therefore developed. It was important to note that super thin glass was superior to polymeric-based substrate because of the inherent properties of glass, including gas barrier properties (which protect against gas and water), heat resistance, and transparency. Thus, it offered a variety of potential applications, i.e. optical devices, electronic paper, solid secondary batteries, and solar cells. From the structural point of view, the idea of super thin glass involved the deposition of inorganic film on top of polymeric substrate and it resulted in lower water absorption ability and gas permeation [100]. However, due to the variation on display technology, the difference on water absorption ability and the oxygen transmission rate was versatile, as exhibited in Figure 13.

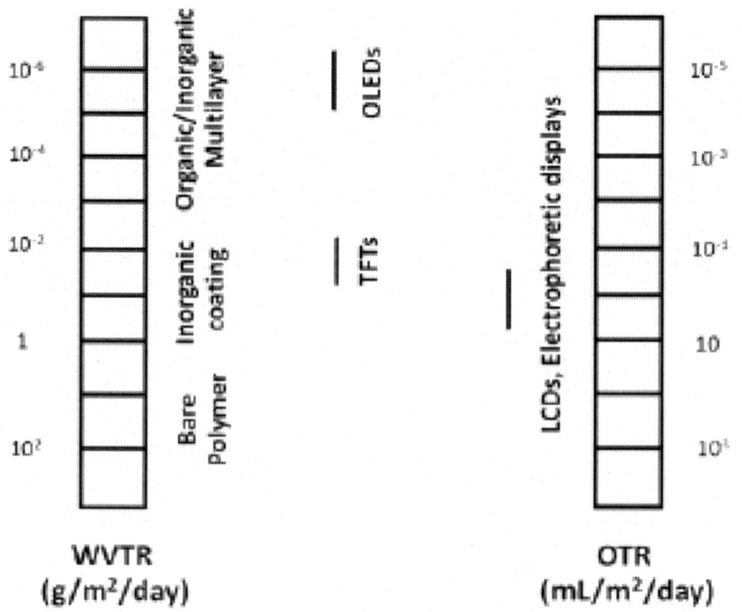

Figure 13. Water vapor transmission rate (WVTR) and Oxygen permeation rate (OTR).

From the fundamental point of view, inorganic transparent oxide films (e.g., silicon and aluminum oxide) were deposited on polymeric-based substrate. They have been extensively employed as gas and water barriers for many electronic devices. It was a major improvement over the water vapor transmission rate (WVTR) [101, 102] and oxygen transmission rates (OTR) [103-105] compared to polymeric-based substrate. The formation of inorganic film was commonly prepared by plasma-enhanced chemical vapor deposition (PECVD), sputtering, or evaporation. However, electronic devices, especially OLEDs, demand a more stringent barrier confinement [106-112]. For an organic light emitting diode lifetime of 410,000 h, the requirements include water vapor transmission rates (WVTR) of 106 g/m^2 /day and OTR of 105 mL/m^2 /day [113].

However, it was important to note that many techniques involving vacuum technology have been employed for film formation. The suitable process for barrier gas was related to plasma enhanced chemical vapor deposition process. It can be employed because of lower temperature in processing polymeric-based substrate.

2.3.1. Device Failure from Barrier Properties

Flexible display technology exhibited many differences compared to previous technologies. It was very sensitive to water and gas and it therefore resulted in degradation of device lifetime and efficiency. Many mechanisms of the effects of water were due to oxidation of anode and cathode. In an organic light emitting diode, metal nano-particles were sometimes employed as electrodes and it was easy to be oxide in the condition of remaining water and air. The formation of metal oxide was therefore obtained.

On the other hand, due to the alteration of organic based conductive materials, it was covered on small molecule and macromolecule, it was easy to detach between water and air to the organic layer, especially on the emissive layer, which was commonly utilized. The transportation of electron and hole was therefore affected and it affected device lifetime and light efficiency. Moreover, if higher remaining oxygen gas existed, it therefore led to oxygen-activated photochemical damage. The degradation of the polymer then occurred due to the existence of water and air [114-116].

Due to the ease of water and gas adsorption, it resulted in dark spot formation on the organic light emitting diode. It is important to note that the dark spot was created from the exposure of water and gas at the interface region in between the conductive layers and it may have occurred at the interface between substrate and conductive layer. To avoid dark spots, many encapsulation techniques were then investigated. Numerous film formation techniques were then employed.

2.3.2. Theories of Gas Permeation

The development and improvement of the quality of substrate from any environmental effect were extensively investigated for prolonging the quality of the organic light emitting diode and its efficiency. Understanding the transport of gas and water should be considered.

From the theoretical point of view, it was remarkable to note that the transportation of gas and water through the polymeric-based substrate can greatly influence the quality and lifetime of the organic light emitting diode. The oxygen ingress into the polymeric-based substrate can cause oxidation in the conductive layer, which led to decreasing quality of the device. On the other hand, water vapor transmission to the polymeric-based substrate can cause moisture, leading to detachment of the conductive layer. As a result, it can cause the undesirable textures in the device structure.

Before understanding the role of barrier technology against moisture and air, it was remarkable to note the difference between gas and vapor. Although these two terms were frequently used interchangeably, the mechanism and its effect still played an important role on device quality, efficiency and lifetime. A gas referred to low boiling point molecules that were non-condensable and exist only in the gaseous state at ambient temperature and atmospheric pressure. Examples of gas commonly detected in polymeric-based substrate are oxygen, nitrogen, hydrogen, and carbon dioxide. The behavior of gas obeyed Henry's law [117-120]. The permeability and diffusivity were constant. However, on the other hand, vapor was theoretically defined as a condensable feature at ambient temperature. It was remarkable to note that in case of vapor, the permeability and diffusivity were concentration dependent for each polymeric-based substrate.

The definition of permeation and gas barriers were closely related. Permeation was inversely proportional to the gas barrier. In case of polymeric-based substrate for flexible display technology, slow rates of gas permeation should be preferably controlled. Figure 14 exhibits the permeation through a polymeric-based substrate. It can be noted that the permeation of a permeant through a polymeric-based substrate was driven by concentration gradient in the direction from high to low concentration.

From Figure 14, the mechanism of permeation consisted of three sequential steps. The first step involved adsorption of the permeant onto the high concentration side of the film surface. After that, diffusion of the permeant across the polymeric-based substrate. Later, the desorption of the permeant from the low concentration side of the film surface.

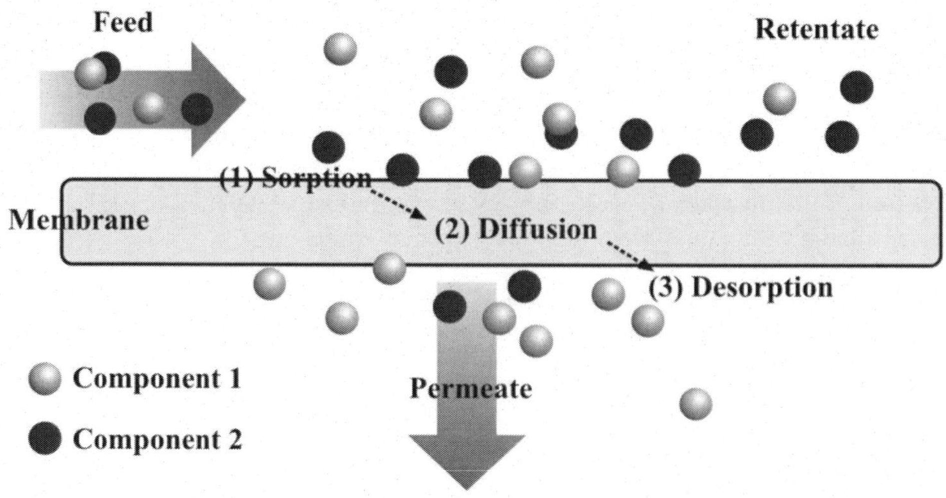

Figure 14. Permeation though a polymeric-based substrate.

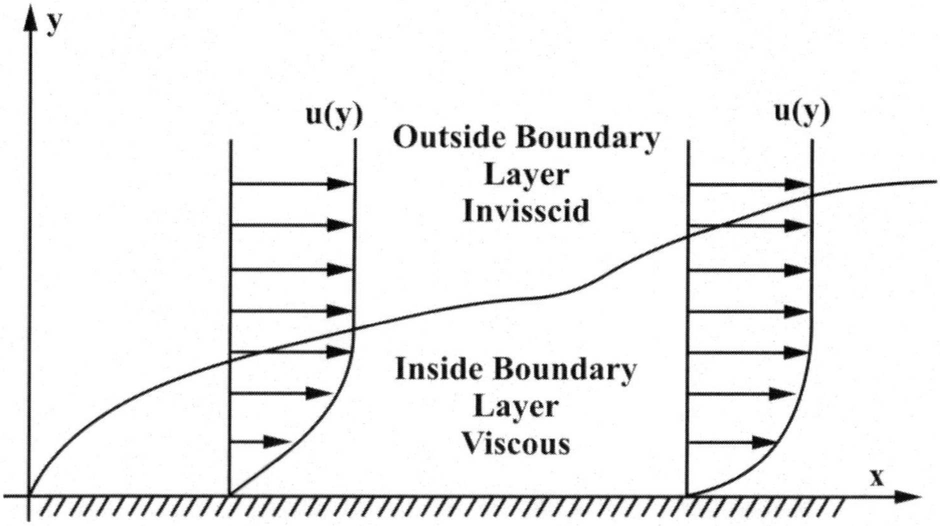

Figure 15. Outside and inside boundary layer of fluid.

Moreover, it can be noted that the permeation rate was therefore conducted; the experiments were usually set up so that both sides of the polymeric-based substrate were exposed to the gas phase. From the diagram, it can be noted that the measurement involved only the gas and solid phase. The permeant in one gas phase entered the solid phase of the polymeric-based substrate and existed in another gas phase.

From the fundamental point of view, the fluid velocity can be categorized into two different regions. One can be controlled and it is called inside boundary layer viscous. The other one cannot be controlled and it is called outside boundary layer in-viscid. The movement of fluid and control on its velocity can be explained in Figure 15 as follows.

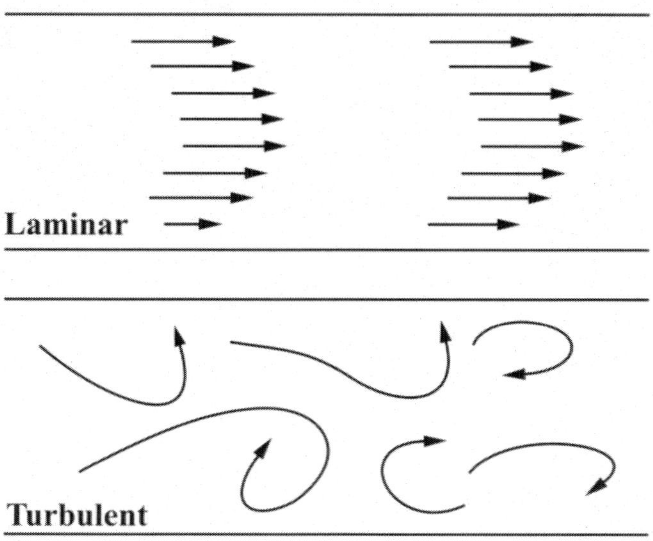

Figure 16. Laminar and turbulent flow characteristic.

Due to the permeation of gas related to fluid flow behavior, it was important to note the flow characteristic. The flow behavior of fluid can be categorized into the regime of laminar and turbulent flow. From the fundamental point of view, laminar flow was considered a well-ordered flow and was characterized by the smooth sliding of adjacent fluid layers over one another, with mixing between layers occurring only on a molecular level. On the other hand, turbulent flow was considered as chaotic flow and was characterized by the transfer of small packets of fluid particles between layers. The flow characteristic of laminar and turbulent is exhibited in Figure 16.

Moreover, in between the laminar and turbulent flow, it can be categorized by Reynolds number [121]. It was the correlation between velocity, inner diameter, fluid density and fluid viscosity. The equation of the Reynolds number can be described as follows

$$Re = D\rho\gamma / \mu$$

Where γ is velocity of fluid, ρ is density of fluid, μ is fluid viscosity and D is inner diameter. Reynolds number was employed to indicate the status of fluid flow. It was unit-less. If the Reynolds number was estimated to be 2000, it was therefore classified as a laminar region. When the Reynolds number was estimated to be 2500, it was classified as turbulent region. The figure in between was a transition region. In order to evaluate the flow characteristic of gas permeation, the Reynolds number should be well designed according to the correlation of velocity of gas, density of gas, as well as its controlled volume. Figure 17 exhibits the transition region. It can be presented in between laminar flow and turbulent flow [122-125].

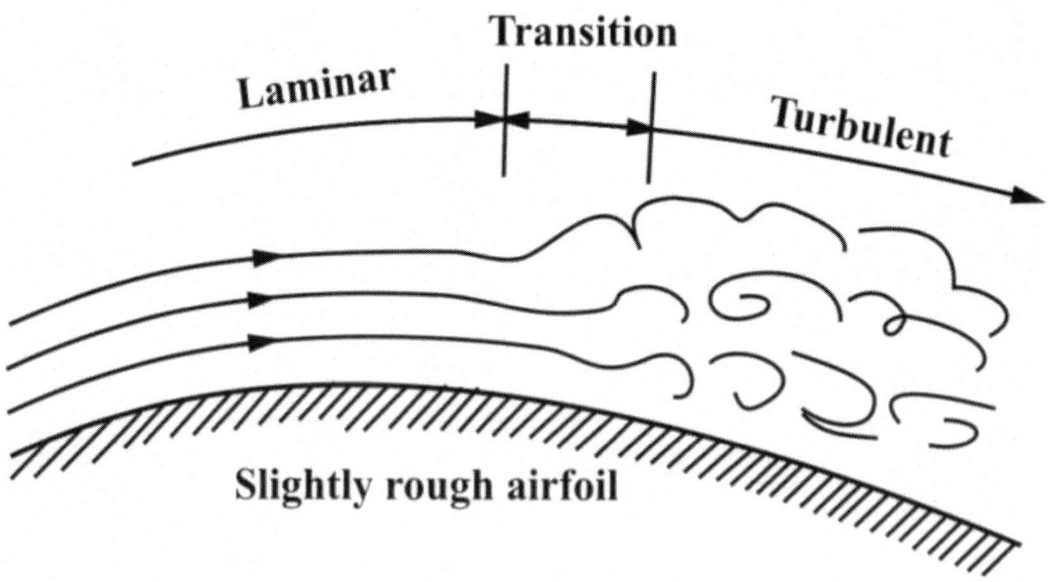

Figure 17. Role of transition region.

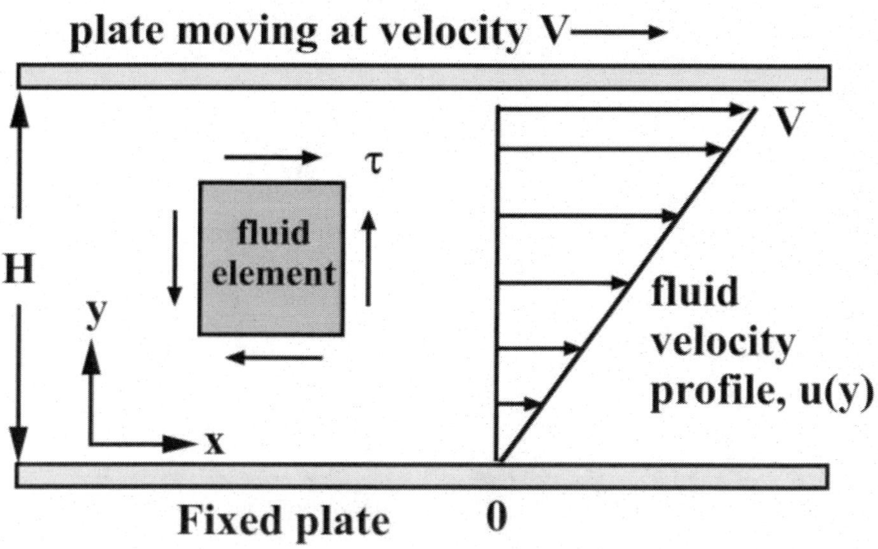

Figure 18. Example of velocity profile of fluid.

Further information on fluid flow can be explained in Figure 18. It was remarkable to note that the upper layer has a velocity V and the lower layer has a slightly smaller velocity. It can be noted that the upper layer was faster; it tended to exert a shear stress on the lower plate in the y direction. In Figure 18, fluid was contained between two large parallel plates separated by a small distance. The upper plate was pulled at a constant velocity in the z direction, whereas the upper plate was stationary. A steady velocity profile was established in the fluid, where v is the fluid velocity in the z direction. H was considered as the vertical distance in between plates. The relationship between shear stress and fluid properties was defined as follows.

$$\tau_{yz} = - \mu \, dV_z/dy$$

This equation was ascribed as Newton's law of viscosity of fluid for one-dimensional flow in the z direction. It can be explained that the minus sign in the equation was due to the fact that the velocity gradient dV_z/d_y was negative. Moreover, in reality, in order to evaluate the movement of fluid, Newton's law of viscosity can be considered in three dimensions and each component has three gradients. The movement of fluid in three dimensions was called a stress tensor for rectangular coordinates. The explanation on stress tensor for rectangular coordinates is exhibited in Figure 20.

The planes and directions associated with these stress components are exhibited in Figure 19. It can be noted that a stress component points in one direction on one surface of the cube and in the opposite direction on the opposite surface. Furthermore, it can be noted that many stress components have been described in three different forms due to the dimension of the controlled volume. The stress tensor and its coordination are indicated in Figure 20.

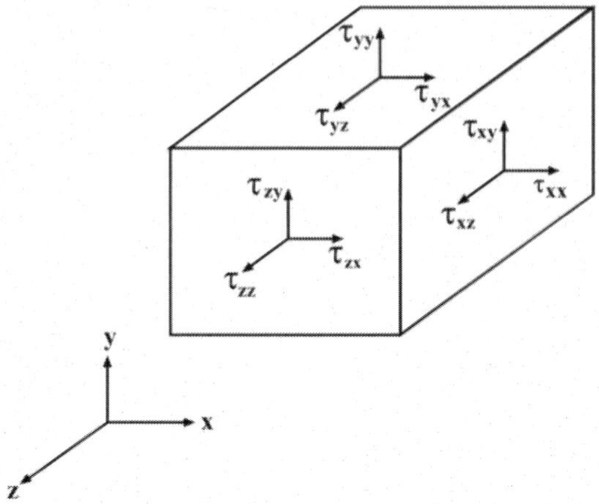

Figure 19. Stress tensor for rectangular coordinates.

Cartesian coordinates (x, y, z):

$$\Delta_{xx} = 2\frac{\partial u_x}{\partial x} \qquad \Delta_{xy} = \Delta_{yx} = \frac{\partial u_x}{\partial y} + \frac{\partial u_y}{\partial x}$$

$$\Delta_{yy} = 2\frac{\partial u_y}{\partial y} \qquad \Delta_{xz} = \Delta_{zx} = \frac{\partial u_x}{\partial z} + \frac{\partial u_z}{\partial x}$$

$$\Delta_{zz} = 2\frac{\partial u_z}{\partial z} \qquad \Delta_{yz} = \Delta_{zy} = \frac{\partial u_y}{\partial z} + \frac{\partial u_z}{\partial y}$$

Cylindrical coordinates (r, θ, z):

$$\Delta_{rr} = 2\frac{\partial u_r}{\partial r} \qquad \Delta_{r\theta} = \Delta_{\theta r} = r\frac{\partial}{\partial r}\left(\frac{u_\theta}{r}\right) + \frac{1}{r}\frac{\partial u_r}{\partial \theta}$$

$$\Delta_{\theta\theta} = 2\left(\frac{1}{r}\frac{\partial u_\theta}{\partial \theta} + \frac{u_r}{r}\right) \qquad \Delta_{\theta z} = \Delta_{z\theta} = \frac{\partial u_\theta}{\partial z} + \frac{1}{r}\frac{\partial u_z}{\partial \theta}$$

$$\Delta_{zz} = 2\frac{\partial u_z}{\partial z} \qquad \Delta_{zr} = \Delta_{rz} = \frac{\partial u_z}{\partial r} + \frac{\partial r_r}{\partial z}$$

Spherical coordinates (r, θ, ϕ)

$$\Delta_{rr} = 2\frac{\partial u_r}{\partial r} \qquad \Delta_{\theta r} = \Delta_{r\theta} = r\frac{\partial}{\partial r}\left(\frac{u_\theta}{r}\right) + \frac{1}{r}\frac{\partial u_r}{\partial \theta}$$

$$\Delta_{\theta\theta} = 2\left(\frac{1}{r}\frac{\partial u_\theta}{\partial \theta} + \frac{u_r}{r}\right) \qquad \Delta_{\phi\theta} = \Delta_{\theta\phi} = \frac{\sin\theta}{r}\frac{\partial}{\partial \theta}\left(\frac{u_\phi}{\sin\theta}\right) + \frac{1}{r\sin\theta}\frac{\partial u_\theta}{\partial \phi}$$

$$\Delta_{\phi\phi} = 2\left(\frac{1}{r\sin\theta}\frac{\partial u_\phi}{\partial \phi} + \frac{u_r}{r} + \frac{u_\theta \cot\theta}{r}\right) \qquad \Delta_{r\phi} = \Delta_{\phi r} = \frac{1}{r\sin\theta}\frac{\partial u_r}{\partial \phi} + r\frac{\partial}{\partial r}\left(\frac{u_\phi}{r}\right)$$

Figure 20. Stress tensor of fluid in rectangular, cylindrical and spherical component.

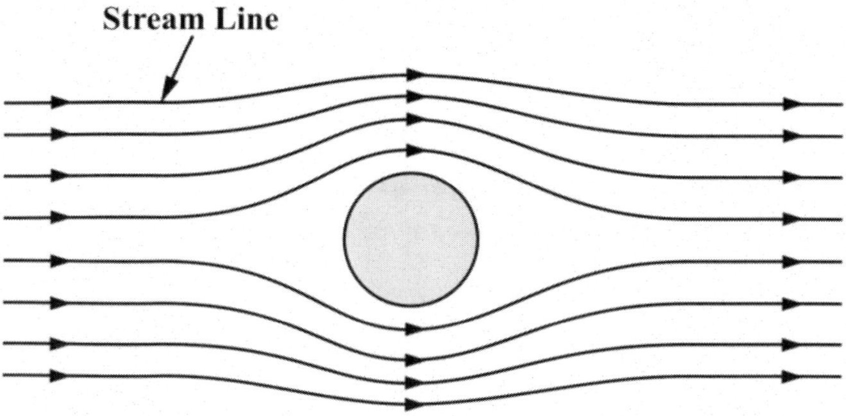

Figure 21. Streamline of fluid flow.

On the other hand, due to the movement of fluid flow, it was involved in streamline flow. From the fundamental point of view, streamline was considered a family of curves showing the pattern of fluid flow, which was drawn in such manner that the velocity vectors for all points along the curves would meet them tangentially [126-128]. Figure 21 exhibits the role of streamline, which was involved in the movement of fluid.

Moreover, the movement of streamline was involved in a tangent of velocity in y and x direction. For mathematics, the tangent between velocity was due to

$$Vy/ Vx = dy / dx = \text{slope of streamline}$$

Where Vx and Vy were considered as the x component and y component of the velocity vector v, respectively. As such, the velocity vector v was tangent to the streamline. Therefore, in three dimensions, the relation becomes

$$dx/Vx = dy /Vy = dz /Vz$$

In steady state, a streamline was considered a path line, which was considered as a line describing the path or trajectory of a given fluid element. In fluid flow, the use of the path line and its experiment is called flow visualization. A tracer, such as dye or smoke, was used to reveal streamlines in a transparent fluid. Up to the present time, the role of streamline was useful for fluid velocity determination. The convergences of streamline represented accelerations, whereas divergence represented deceleration. Since the cross-sectional area available for flow was smaller downstream than upstream, the average fluid velocity had to be higher downstream than upstream. This acceleration of fluid was reflected by the convergence of the streamlines at the contraction.

The role of gas permeation for flexible display substrate was considered one of issues that we need to revise. One of mechanisms was related to mass transfer, which involved concentration gradient. The role of mass transfer can be either by diffusion or conventional process. From the fundamental point of view, diffusion was considered as the mass transfer that occurred across a stationary solid of fluid in which the concentration gradient existed. In contrast, convection referred to mass transfer that occurred across a moving fluid in which a concentration gradient existed. The understanding of concentration gradient can be further explained in Figure 22. Between different concentration regions, the diffusion primarily occurred, subsequently followed by convection. No difference can be noted in concentration gradient, no mass transfer.

Up to the present time, the role of gas permeation provided many effects on device quality. To use flexible technology with higher efficiency, many developments on the barrier technique should be considered.

3.3. Barrier Coatings on Polymer Substrates

Due to the effect of gas permeation on device lifetime and quality, many efforts have been extensively developed for prolonging the lifetime of the device as well as to maintain its quality. Since the flexible substrate involved transparency, a barrier film against moisture and gas should be considered on its transparency as well.

In the past, transparent glass sheet was used as substrate and the ability for water absorption of glass was very low. This caused much interest in the organic light emitting diode. However, novel display technology changed and therefore the use of glass sheet as substrate was limited, if flexible properties of substrate were required. However, to gain the beneficial surface properties of flat glass sheet, it was important to investigate the structural properties of flat glass sheet. From the fundamental point of view, it was important to note that the structure of glass was due to ceramic oxide. The example of ceramic oxide was due to silicon oxide, silicon nitride siliconeoxynitride, aluminum oxide and the mixture of these [129-134]. It was important to note that compared to polymeric-based substrate, the water absorption ability of ceramic was much less and it rendered a longer lifetime of the electronic device. Up to the present time, due to the significant effort of ceramic-based materials for substrate of display technology, it helped bring these properties to the surface of flexible substrate. This challenge involved the development of polymeric-based substrate through the growth of ceramic oxide layer. However, due to the nature of ceramic, which was commonly recognized as a fragile and brittle feature, the growth of the ceramic layer on the polymeric substrate should be considered mechanical properties. Care should be used when correlating the ceramic oxide layer on top of polymeric substrate.

Up to the present time, in order to gain the excellent optimization of the use of polymeric substrate with the growth of ceramic oxide layer, many investigations have been performed [135-137]. The challenge caused the growth of ceramic oxide layer in a few nanometers thickness, which reduced the diffusion of water and oxygen to acceptable levels. Moreover, many approaches were involved regarding the structure and the number of barrier layers, which were called dyads. It involved a hybrid of alternative layers with organic and inorganic materials. A typical single barrier layer provided only two to three orders of magnitude improvement in the barrier properties compared to a bare polymer substrate. To satisfy the requirements of OLEDs, polymer substrates must be coated with multilayer organic–inorganic thin films.

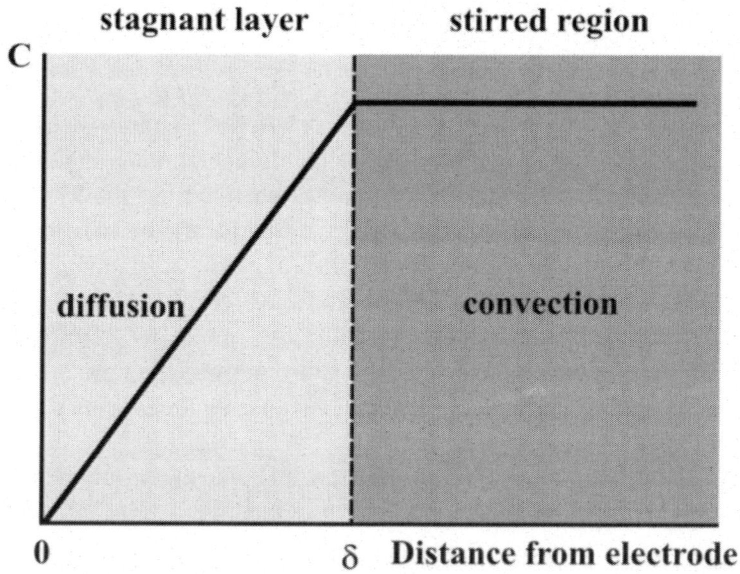

Figure 22. Mass transfer: Diffusion and convection.

Moreover, there is an alternative method involving the ceramic oxide film formation. From a fundamental point of view, polymeric-based substrate for flexible display technology exhibited limitations due to thermal resistance. The use of a high temperature process for ceramic oxide growth was therefore limited in processing polymeric-based substrate. The limitation was due to the effects of temperature. It was commonly known that the quality of the inorganic film was detrimental for processing using low temperature. There has been much research dedicated to finding a way to obtain high-quality films at low temperatures by developing new deposition technologies and improving the deposition processes, including sputter deposition, electron beam deposition, and plasma enhanced chemical vapor deposition. These techniques were interesting for the production of ceramic oxide layer on top of the polymeric-based substrate due to the quality of film formation. However, the use of these techniques was still limited in the mass production process, which was commonly employed in industrial commercialization. The use of vacuum technology such as sputter deposition, electron beam deposition and plasma enhanced chemical vapor deposition was still limited in mass production. Moreover, the use of these techniques was also limited because of safety and economical concerns. From the viewpoint of industrial commercialization, processing should be available for in-line production through a roll-to-toll process (RTR) including polymeric-based substrate formation, barrier technology against moisture and gas as well as any active layer for device deposition. In industrial commercialization, Vitex System developed "Flexible Glass," which was a commercially available, polymeric-based substrate utilizing defect-decoupling layers of vacuum deposited. The example caused the case of poly acrylate and its barrier, which is composed of multiple layers of gas barrier materials such as silicon oxide. This substrate was sufficiently impermeable to moisture and oxygen for application in organic light emitting diodes with a low water vapor transmission rate.

3.4. PERMEATION RATE MEASUREMENTS

The organic light emitting diode has undergone significant development, but there were no standards for the measurement of barrier properties. Highly sensitive permeation measurements were crucial for the characterization and development of polymeric substrates for flexible display applications, especially organic light emitting diodes, which were very sensitive to water and oxygen. The current specifications for the permeation rates of organic light emitting diode were on the order of 10^5 mL/m^2 /day at standard temperatures and pressure for oxygen and 10^6 g/m2 /day for water vapor. However, there is no commercial method available for the measurement of ultralow moisture permeation for OLED applications. Therefore, the development of a moisture permeation measurement system will be essential for investigating permeation mechanisms and understanding device lifetime and degradation phenomena.

The commercial equipment available from MOCON was limited to a sensitivity of 5-10^4 g/m^2/day under varying temperature conditions. Researchers have developed a method based on detecting the optical degradation of calcium to measure the permeation rate of organic light emitting diodes. This calcium test method can effectively increase the speed of research and development of thin-film encapsulate suitable for OLED devices by providing quantitative results and sample configurations inaccessible to commercial equipment, which is now limited to 5-10^3 mL/m^2 /day for oxygen and 5-10^3 g/m^2 /day for water for 4-in diameter samples of coated polymeric-based substrate. Effective water vapor transmission rates down to 10^6 g/m^2 /day were detected. Many researchers therefore proposed the idea of measuring the amount of oxidative degradation in a thin calcium sensor by in situ resistance measurements. The resulting baseline contribution to the water vapor transmission rate of a glass reference is below 10^6 g/m^2 /day under accelerated test conditions.

Chapter 3

Overview of Cellulose: From Forestry to Industrial Commercialization

3.1. Overview of Cellulose: From Forestry to Industrial Commercialization

Cellulose, the most common bio-polymer, is a raw material from trees and other plants that has been used for centuries in various applications [53, 138-140]. It was first isolated from plant matter by French chemist Anselme Payen in 1839. He reported that cellulose has a structure identical to starch, but it exhibited a difference in structure and properties. Up to now, cellulose is commonly known as a polysaccharide with the common formula $(C_6H_{10}O_5)_n$, and consisting of a linear chain of several hundreds to over a thousand linked glucose units.

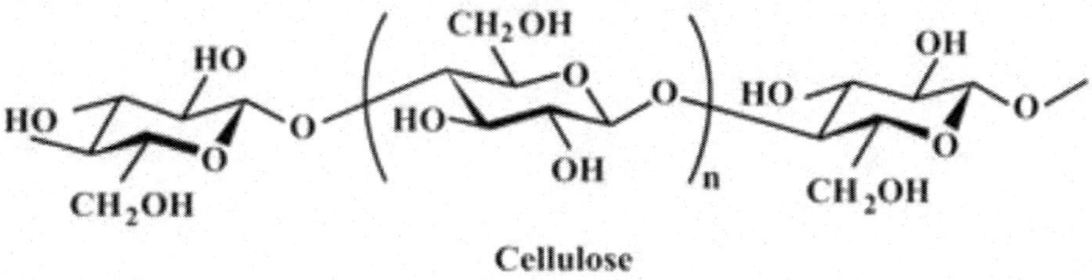

Figure 23. Chemical structure of cellulose.

Figure 23 exhibits the chemical structure of cellulose. Cellulose is composed of several glucose units in order to form long linear chains, so-called "polysaccharides." Cellulose can be produced from plants or bacteria. Cellulose is extracted from plant materials such as wood, flax, hemp, sisal or cotton. For plants, cellulose is found in a composite form composed of polymers, lignin and carbohydrates such as hemicelluloses and cellulose, which are physically and chemically bound together. The matrix, which is lignin in plant matter, binds the cellulose fibers together acting as a resin system [141-143]. Cellulose and hemicellulose act as reinforcement materials and an interfacial compatibilizer between cellulose and lignin, respectively [144-146]. Although cellulose constitutes a significant portion of material, mechanical properties such as tensile strength and stiffness of the plain material are inferior to the properties of neat cellulose. When comparing to natural fibers, it can be observed that mechanical properties, such as tensile strength and Young's modulus, commonly drop with lower cellulose content. For high strength materials, it would be preferable to have as much cellulose in a natural material as possible.

Table 1. Composition and comparative properties of natural and man-made fibers

Types of fiber	Cellulose (%)	Lignin (%)	Hemicellulose (%)	Pectin (%)	Wax (%)	Tensile strength (MPa)	Young's modulus (GPa)
Fruit							
Coir	36-43	41-45	0.15-0.25	3.4	-	131-175	4-6
Bast							
Jute	61-71.5	12-13	13.6-20.4	0.2	0.5	393-773	13-26.5
Flax	71	2.2	18.6-20.6	2.3	1.7	345-1100	27.6
Hemp	70.2-74.4	3.7-5.7	17.9-22.4	0.9	0.8	690	-
Ramie	68.6-76.2	0.6-0.7	13.1-16.7	1.9	0.3	400-938	61.4-128
Leaf							
Sisal	67-78	8-11	10.0-14.2	10.0	2.0	468-640	9.4-22.0
PALF	70-82	5-12	-	-	-	413-1627	34.5-82.51
Seed							
Cotton	82.7	-	5.7	-	0.6	287-800	5.5-12.6
Man-Made							
E-glass	-	-	-	-	-	2000-3500	70
S-glass	-	-	-	-	-	4570	86
Aramid	-	-	-	-	-	3000-3150	63-67
Carbon	-	-	-	-	-	4000	230-240

Table 1 exhibits the broad composition and comparative properties of natural and man-made fibers. From the structural point of view, it was important to note that cellulose fibers exhibit a unique structural hierarchy derived from their biological origins. The use of cellulose and modified cellulose can be versatile, depending on specific properties that can therefore refer to relevant application [147-150].

Furthermore, it was remarkable to note that cellulose derived from plants exhibit the hierarchical structure of cellulosic plant-based material. Cellulose can be effectively prepared from wood and cell walls of wood. Physical and chemical modifications were provided as modified cellulose. The investigation on the structure and properties of cellulose has been recent. Numerous improvements of cellulose research have been conducted by using it as a reinforcing phase in composite material, which is commonly referred to as a combined material. The one, called matrix phase; which is generally referred to as polymer matrix, and the other one, called the reinforcing phase; which refers to cellulose. Some researchers therefore called cellulose based composite material. The reason for the development of cellulose-based composite was due to the excellent properties of cellulose including high stiffness, low co-efficient of thermal expansion, high thermal resistance as well as high chemical resistance. Conventionally, polymeric materials in industry were limited in some applications only if high thermal resistance is required; while neat ceramic materials are also inappropriate if displacement is necessary. The design of composite features can effectively overcome these limitations. Flexibility, and thermal resistance as well as chemical resistance can be designed. In essence, for composite material; cellulose has therefore been prepared on nano-scale forms in order to insert in polymer matrix. Nano-scale form of cellulose can enhance the surface area; which commonly refers to force absorption if the composite will be applied in many engineering applications. Furthermore, it can be noted that prior to using cellulose, it can be categorized by crystallinity into 2 different structures, called nanofibrillated cellulose and nanocellulose whiskers.

The first successful isolation of cellulose microfibrils was reported in last three decades. Some of researcher called cellulose microfibril or microfibrillar cellulose. The application of this form of cellulose strongly related to nano-form of composite, which required a high surface area before insertion in polymer matrix. Generally, nanofibrillated cellulose is considered cellulosic material, composed of expanded high-volume cellulose, moderately degraded and greatly expanded on the surface area. The diameter of this cellulose is in the range of 20-50 nm and the length can be versatile, depending on the physical modification process such as homogenization, the pulping process, as well as stream explosion.

From the past, the feature of nanofibrillated cellulose initially related to softwood. It was hydrolyzed in 2.5 M hydrochloric acid (HCl) at 105°C and afterwards pumped through a slit homogenizer. Nanofibrillated celluloses were prepared from hydrolysis reaction and it subsequently passed through the homogenizer. The resulting gels showed an obvious increase in viscosity after several passes through the homogenizer. The term microfibril is misleading as it does not reflect the real dimensions of the fibril. Moreover, it is not possible to obtain a perfectly homogeneous sample of single cellulose microfibrils. Therefore, mechanical disintegration of pulp usually aims at the isolation of cellulose fibril aggregates having diameters below 100 nm. In this work, such cellulose portions are generally termed nanofibrillated cellulose.

The other type of cellulose, called nano cellulose whiskers, was considered crystallinity of cellulose. Long chains of cellulose were cut into small fragments and bleached, followed by a disintegration process using a blender. SEM images revealed rod-like, highly crystalline cellulose whiskers with diameters and lengths in the nano scale. The use of nano cellulose whiskers was considered in applications in which high crystallinity was required. Cellulose nano-form whiskers can be prepared in many forms such as rod-like, tube-like and spherical shape. The application of this cellulose was also related to cellulosic based materials. The reinforcing effect of cellulose whiskers was compared to the effect of nanofibrillated cellulose in poly (styrene-co-butyl acrylate) latex. It was exhibited that both fillers led to an increase in mechanical properties. However, the incorporation of nanofibrillated cellulose resulted in significantly higher values, due to the entanglements between the fibrils leading to a rigid network of nanofibrillated cellulose.

3.1.1. Cellulose Derived from Plant: Modification

Before the use cellulose in any application, it was remarkable to note that physical and chemical modification techniques were employed to modify the properties of cellulose. The modification technique for cellulose can be categorized into two different methods. One involved pre-treatment. It definitely referred a preliminary step for cellulose modification. The content of pre-treatment involved purification of cellulose derived from plant by chemical and physical methods. The other one involved post-treatment; this step was employed in order to adjust the cellulose for each specific application. Up to the present time, the use of cellulose and its preparation process was versatile. In order to understand the process of cellulose preparation, the fundamental concept of these techniques was explained as follows.

Homogenizer was often used to manufacture microfibrillated cellulose. Using the homogenizer process, cellulose suspension was pumped at high pressure and consequently fed through a spring-loaded valve assembly [151]. The process of open and close valve was very rapid, and fiber was therefore subjected to a large pressure drop under high shearing force. In the experiment, the applied pressure was very high and resulted in high shear rate and it subsequently provided very thin cellulose particles. The size of the chamber of the homogenizer process needed to be carefully controlled in order to obtain uniformity in size of cellulose particles. However, in order to scale up to commercialization, the use of this technique should be considered a high-energy consumption process.

The grinding process was called a grinder. The principle consisted of the breakdown of the cell wall structure owing to the shearing forces generated by the grinding stone with high speed of grinding rotation [152, 153]. Then, cellulose that composed the cell wall in a multilayer structure was therefore individualized from wood. The size of cellulose can be effectively controlled by grinding round and speed. However, there is no significant change in morphological properties.

Cryocrushing is nowadays rarely used in cellulose suspension preparation. This process typically consisted of the crushing of frozen pulp with liquid nitrogen. Ice crystals within the cells were then formed, and under mechanical crushing, they slashed the cellular wall and release wall fragment. Typically, the size of cellulose was 20-40 nm and the length can be varied in several thousand nanometers. It is commonly used to exact cellulose from agriculture crops and by products such as flax and hemp.

Electrospinning is rarely used for cellulose suspension manufacturing, while providing excellence in engineering properties of cellulose. The process was performed by applied electric charge to draw very fine fiber from liquid. This was probably due to the fact that electrospinning effectively combined characteristics of both electrospraying and conventional solution dry spinning of fiber. This process did not require high temperature in manufacturing [154-156]. Cellulose prepared from this technique was very uniform. The controllable size and shape can be effectively monitored depending on applied electric charge.

On the other hand, there have been numerous techniques for modifying the properties of cellulose. An example of chemical modification is explained in detail as follows. Prior to using cellulose suspension with higher efficiency, many strategies have been proposed in order to obtain cellulose that were less stiff and cohesive, therefore a reduction on the energy for cellulose suspension production was required. From a structural point of view, there were three alternative methods; limited hydrogen bond or adding a repulsive charge, and decreasing the degree of polymerization or the amorphous link between individual cellulose particles.

Enzymatic pre-treatment was considered a sub-method after mechanical shearing and high-pressure homogenization. Generally, from the past, the use of acid hydrolysis was performed. However, the use of enzymatic pre-treatment was less aggressive and it consequently allowed for selective hydrolysis for non-crystalline cellulose and it subsequently facilitated the mechanical disintegration. Cellulose was easily extracted from wood within any lignin, pectin and hemicelluloses. The homogenous cellulose suspension can be effectively obtained.

The most effective and common method for pre-treatment was called “TEMPO mediated oxidation” [157-159]. This pre-treatment is the most promising method for effecting the surface modification of native cellulose. The carboxylate and aldehyde functional groups can be introduced into solid native cellulose under aqueous and mild conditions. For the production process, energy consumption can be saved from the mechanical treatment process. In this pre-treatment, the basic principle is to form oxidation of cellulose via the insertion of NaClO to aqueous cellulose suspension in the presence of catalytic amounts of 2,2,6,6-tetramethyl-1-piperidinyloxy (TEMPO) and NaBr at pH 10-11 at ambient temperature. The C6 primary hydroxyl groups of cellulose were therefore selectively converted to carboxylate groups via the C6 aldehyde groups, and only the NaClO and NaOH were consumed. Consequently, the nanofibrils within the fibers separate from each other better due to the repulsive forces among ionized carboxylates, which overwhelm the hydrogen bond holding them together. The more addition of NaClO, the more carboxylate groups was formed on the surface of cellulose. The use of this pre-treatment was effectively conducted for all types of wood including flax and hemp.

Carboxymethylation was one type of effective chemical treatment. The concept of this pre-treatment is to increase the anionic charges in the formation of carboxyl groups on the surface of cellulose [160]. As a result, this chemical treatment made the fibrils highly charges and easier to liberate. However, this technique required high-energy consumption for pre-treatment while passing through a microfluidizer. Due to this reason, the development of acetylation pre-treatment was performed. The grafting of acetyl moieties aimed to decrease the hydrophilic properties and to enhance the chemical affinity between cellulose and a non-polar solvent. Furthermore, this concept was further developed if a cellulose based polymer composite was performed. The grafted acetyl groups reduced the hydrogen bonding between cellulose and therefore allows for superiority on dispersibility in an apolar polymer matrix. This concept provided good efficiency in opening wide-vision for industrial commercialization scale production.

Post-treatment was theoretically considered as the last step before using cellulose with desired properties and performance. From the prospective application, post-treatment was quite specific to consider. Recently, the development of barrier properties against moisture and gas permeation was extensively considered. The modification on the cellulose surface by solvent-exchange and epitaxial thin film growth was employed. On the other hand, in order to fabricate as cellulose based composite, increasing the surface adhesion between the cellulose and the polymer matrix was also considered. The grafting of coupling agent in different amounts was employed.

Apart from natural plants, there are some different species of bacteria that can produce pure cellulose. Certain bacteria belonging to the genera Acetobacter, Agrobacterium,

Alcaligenes, Pseudomonas, Rhizobium or Sarcina were reported capable of producing cellulose, the most efficient producer being Acetobacter xylinum. Bacterial cellulose is chemically the same as plant cellulose, which is β-1,4-glucans.

Further information on bacterial cellulose was written about in a scientific paper by Brown in 1886. It is chemically the same as plant cellulose, which is β-1,4-glucans. The papers described a fermentative process, which formed a gelatinous transparent membrane at the surface of an acetic fermentation. The membrane had the capability to grow to a thickness of 25 mm and proved to be very strong and tough. It also called pellicle, proved to be cellulose formed by a bacterium. The bacterium was named Bacterium Xylinum by Brown, but later it was classified as Acetobacter Xylinum [161-163]. However, in general, certain bacteria belonging to the genera *Acetobacter*, *Agrobacterium*, *Alcaligenes*, *Pseudomonas*, *Rhizobium* or *Sarcina* were reported capable of producing bacterial cellulose, the most efficient producer being *Acetobacter xylinum*. In Table 2, it exhibits a classification of bacteria's ability to produce cellulose.

Acetobacter xylinum is a strain of acetic acid-producing bacteria and an obligate aerobe. Cellulose is a hydrophilic glucan polymer, which contains hydroxyl groups. These hydroxyl groups can form intramolecular hydrogen bonds inside the macromolecule itself and among other cellulose macromolecules as well as with hydroxyl groups from the environment. In the bacterial culture, the bacterial cellulose microfibrils agglomerate and form a pellicle, which will provide a firm matrix that floats and, therefore, allows the embedded bacteria to stay in close contact with the atmosphere. The produced cellulose pellicles play an important role in promoting colonization of the cells on the substrate and provide protection against competitors. Cellulose pellicles were also observed to protect *Acetobacter xylinum* cells from UV light.

Bacterial cellulose has been substantially utilized in the food industry [60]. An example is nata de coco, a coconut gel dessert prepared by bacterial fermentation of coconut water. Though originally a Filipino dessert, it became a very popular diet food in Japan. Also, owing to its good liquid adsorption, ability to withstand sterilization and non-allergenic properties, bacterial cellulose is extensively employed in many medical applications. The examples are artificial skin, blood vessels and ureters. Bacterial cellulose is also applied as a binder in the paper industry to improve the strength as well as the durability of paper. Other uses include diaphragms in audio components. The growth of *Acetobacter xylinum* and its production of bacterial cellulose are governed by its carbon source, nitrogen source and the condition of the culture system [164-166]. Many carbon sources were studied as reported in the literature. Many researchers have extensively investigated the effect of carbon sources on the production of bacterial cellulose. It was found that galactose and xylose gave lower yields than glucose due to slower growth rates of the bacteria. Moreover, it can be noted that the bacterial cellulose yield from sucrose was only half the yield from glucose whilst fructose and glycerol gave almost the same cellulose yield as glucose. The sugar alcohol, arabitol and mannitol, were reported to yield 6.2 and 3.8 times more cellulose than glucose. It was shown that these sugar alcohols were metabolized via xylulose and fructose and that no gluconic acid was produced during fermentation, which consequently resulted in more stable pH of the culture. Other carbon sources, such as 5- or 6-carbon monosaccharides, oligossaccharides, starch, alcohol and organic acid were also reported as well as the methylated glucose derivative.

Acetobacter xylinum also requires a specific complex nitrogen source such as yeast extract and peptone for its growth. Amino acids such as methionine and glutamate are claimed to have an important effect on the cell growth and the cellulose production, while pyridoxine, nicotinic acid, p-aminobenzoic acid and biotin are the most stimulating vitamins. However, one experiment did not observe any positive effect on cellulose synthesis of vitamins. Some other compounds that can strongly stimulate cellulose production include derivatives of choline, betaine and fatty acids. Corn steep liquor, which is a viscous concentrate of corn solubles that is rich in vitamins, amino acids, minerals and other growth stimulants, is also found to be suitable for cellulose production by *Acetobacter xylinum* strain *Gluconacetobacter xylinus* BPR 2001. It was reported that when lactate, which was detected in corn steep liquor, was added to culture media containing other nitrogen sources, the bacterial cellulose production was enhanced to the level similar to those in the medium containing corn steep liquor. In terms of the bacterial culture system, studies indicate that the optimal pH range for the production of bacterial cellulose by *Acetobacter xylinum* is in the range of 4–6 while the optimal growth temperature is in the range of 25–30°C. Variation in temperature can cause changes to the degree of polymerization of the cellulose and its water-binding capacity. Oxygen transfer is particularly crucial to bacterial cellulose production. It has been found that when oxygen-enriched air was supplied into the system, it supplied culture comparable to that in a mechanically agitated stirred-tank fermenter, but with much lower energy consumption. The size of the cellulose pellets also strongly depends on the amount of dissolved oxygen in the broth. Bacterial cellulose formed under a low oxygen environment has relatively fewer ramifications compared to those formed under a higher oxygen environment. However, high levels of operating pressure (oxygen and other gas pressures) have been reported to inhibit the bacterial growth.

Table 2. A classification of the bacteria's capability to produce cellulose

Genus	Cellulose structure
Acetobacter	*pellicle composed of ribbons*
Achromobacter	*fibrils*
Aerobacter	*fibrils*
Agrobacterium	*short fibrils*
Alcaligenes	*fibrils*
Pseudomonas	*no distinct fibrils*
Rhizobium	*short fibrils*
Sarcina	*amorphous cellulose*
Zoogloea	*not well defined*

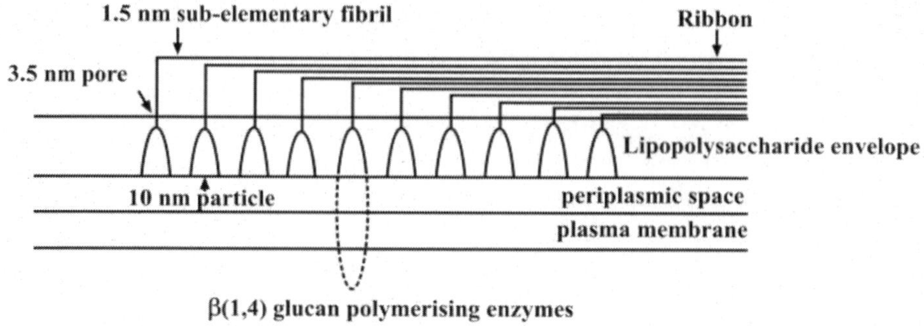

Figure 24. Scheme for the formation of bacterial cellulose.

Moreover, the use of bacterial cellulose was very versatile; many technological approaches have been developed on its quality and feasibility to scale-up for mass production process [167-171]. From the past, the conventional method for the production of bacterial cellulose is static cultivation using sugars such as glucose, fructose and sucrose. However, this method is not applicable to large-scale industrial production due to the long cultivation time and expensive carbon source. However, it was later shown that it is difficult to obtain good quality pellicles, and the rotating system is not successful enough for more intensive studies for industrial purposes. Bacterial cellulose production in stirred-tank reactors and aerated reactors has also been reported. In terms of the cellulose quality, the bacterial cellulose cultured in the static system exhibits a more compact and denser structure than those cultured in agitated systems. It also has a much higher degree of polymerization, crystallinity index, and Young's modulus. In addition, the agitated systems were reported to offer lower productivity than the static system due to their spontaneous generation of cellulose-negative mutants, which do not produce cellulose. In general, the structure of bacterial cellulose reported mostly uniaxial oriented ribbons cellulose, which was formed in the static culture, whereas cellulose synthesized under agitating conditions demonstrated a structure of disorderly, curved and overlapping ribbons. Such a disordered microstructure could result from constant shear forces during the agitation.

The synthesis of cellulose in *Acetobacter xylinum* occurs between the outer cell membrane and cytoplasma membrane by a cellulose-synthesizing complex, which is associated with pores at the surface of the bacteria. The cellulose synthase is considered the most important enzyme in this process. The cellulose produced leaves the pores as fibrils and they join many other fibrils to form a ribbon of crystalline cellulose. The ribbon elongates in direct association with the cell envelope and remains associated in the cell. The self-assembly process should be responsible for the crystallization of the obtained cellulose. Some authors suggested that the process of cellulose formation should occur at the medium/pellicle interface and that the producing bacteria should be near this interface. It demonstrated that the real site for the cellulose formation was at the upper side of the pellicle at the air/pellicle interface, which meant that the producing cells as well as nutrition must be transported through the pellicle to its surface. A principal biochemical pathway from glucose to cellulose is shown in Figure 24. A single *Acetobacter xylinum* was reported as being capable to polymerize up to 200,000 glucose molecules per second into a cellulose chain. Apart from the cellulose, other by-products are also synthesized such as carbon dioxide, acetan, which is a branched polymer of β-1,4-glucose, acetic acid, and gluconic acid.

Table 3. Properties of bacterial cellulose

	Features	
	Agitated Culture	Static Culture
Crystallinity index (%)	63	71
Crystallite size (nm)	6.9	7.4
Crystallinity (%)	72	80
Cellullose I (%)	61	73
Degree of polymerization	10900	14400

Bacterial cellulose synthesized by *Acetobacter xylinum* has been extensively investigated. Some of the reported properties are listed in Table 3. An air-dried bacterial cellulose sheet has tensile strength of 256 MPa with the Young's modulus of 17 GPa. The elastic modulus of bacterial cellulose single fibril recently measured using atomic force microscopy (AFM) was 78 GPa, which is comparable to glass fiber. The elastic modulus of the bacterial cellulose single fibril, also estimated using a Raman spectroscopic technique was 114 [61]. This modulus was predicted from a calibration of the Raman band shift against the elastic modulus. In comparison, the elastic modulus of the crystalline cellulose was measured by X-ray Diffraction (XRD) to be 138 GPa as reported.

The main advantage of bacterial cellulose is its biodegradability and environmentally friendly aspect. In the last few years, various technological innovations have been pressed by the world's concern with the depletion of natural resources and the impact of technology on the environment and climate change. Ultimately, there is a strong focus on renewable and biodegradable materials. An important issue with most natural materials used nowadays refers to the mechanical properties, which are far more inferior to the properties of synthetic materials such as glass, carbon fiber and aramid used in various different applications, due to their high performance and versatility. Bacterial cellulose is one of the innovative materials capable of incorporating the renewability and biodegradability of natural materials with high mechanical properties of glass fibers.

Other advantages of bacterial cellulose refer to the processing step: the wear on milling tools is almost negligible compared to that on tools used for the fabrication of parts of metal or synthetic composite parts. Compared to glass, it has a lower density and favorable energy absorption, good acoustic and thermal properties. It offers higher toughness than glass, which means it does not splinter like glass fiber. It has the potential for one-step manufacturing even of complex construction elements and is capable of being made almost completely transparent. Health risks that come with splintering of the material or use of toxic materials are much smaller when working with bacterial cellulose. In addition, production of metals, glass and carbon fibers requires very high production temperatures as well as a lot of energy, which also dominates the price.

One of the most attractive characteristics of bacterial cellulose for researchers is the fact that bacterial cellulose can be produced in predetermined shapes. Because bacterial cellulose is formed on the interface between air and culture medium, the shape of the cellulose is directly determined by the shape of the interface medium.

3.2. Modification of Cellulose as Composite Based Material

Cellulose reinforced composite materials consist of bacterial cellulose of high strength and modulus embedded in or bonded to a matrix with distinct interfaces (boundary) between them. In this form, both bacterial cellulose and matrix retain their physical and chemical identities, yet they produce a combination of properties that cannot be achieved with either of the constituents acting alone [52, 148, 150]. In general, bacterial celluloses are the principal load-carrying constituent while the surrounding matrix keeps the bacterial celluloses in the desired location and orientation, and acts as a load transfer medium between them [172-175]. The matrix also protects the reinforcing bacterial celluloses from environmental attack due to temperature and humidity, for example. Thus although the bacterial celluloses provide reinforcement for the matrix, the matrix also serves a number of useful functions in composite materials. The mechanism of reinforcement can be explained as the transfer of the stress applied to the composite to the reinforcing bacterial celluloses by the deformation of the matrix material around the bacterial celluloses. Assuming that a good bond is maintained between the bacterial celluloses and the matrix, when stress is applied to the composite, the bacterial celluloses will resist strain locally and a much higher stress will be induced in the bacterial celluloses than in the matrix around them. The matrix around the bacterial celluloses will be constrained and will suffer shear at the area near the bacterial cellulose tips. This shear of the matrix is the dominant process in determining the composites properties. To illustrate this fact it is only necessary to imagine a deformable sheet on which two sets of equally spaced lines are drawn perpendicular to one another. This construction gives a regular pattern of identical squares. If the sheet is uniformly stretched in the direction of one of the sets of lines, the squares become rectangles, retaining the 90° angle in each corner of the original squares. However if a bacterial cellulose is embedded into the original un-deformed sheet (matrix), as the sheet is stretched, the results can be illustrated in Figure 25. Part of the sheet that is adjacent to the bacterial cellulose is obliged to follow the deformation of the stiffer bacterial cellulose whilst further away the sheet can deform to a larger extent. The result is that there are no longer intersections of the lines at right angles. The effect will be greatest in the vicinity of the bacterial cellulose tips [176-178]. The presence of the bacterial cellulose has caused a stress concentration in the area near the two bacterial cellulose tips and much lower stress around the [179] body of the bacterial cellulose. However this stress concentration at the tips can be reduced if the stress can be transferred to the adjacent bacterial cellulose. Therefore, reinforcing bacterial celluloses must be close enough to one another in order to maximize the reinforcing effect, which can be achieved by an acceptable dispersion level of bacterial celluloses into the matrix.

The properties of composites are strongly influenced by the proportions and properties of the matrix and the reinforcement. Most properties of a composite are a function of a number of parameters as the constituents usually interact in a synergistic way. The strength of the interface between the bacterial cellulose and the matrix is particularly important in determining the properties of the composites. The interfacial bond strength has to be sufficient for load to be transferred from the matrix to the bacterial celluloses if the composite is to be stronger than the unreinforced matrix. Other parameters, which may significantly affect the properties of a composite are the shape, size, orientation and distribution of the reinforcement within the matrix as well as various features of the matrix such as the crystallinity and the grain size for crystalline polymer matrix [180].

Generally, cellulose composite has high water absorption behavior. This characteristic must be considered because it will interrupt and decrease the performance of cellulose composite. In general, water absorption can be determined by two methods: one – the cellulose composite has porous structure, two – the cellulose can form hydrogen bonding with water. It is commonly known that cellulose is insoluble in water; it can form hydrogen bonds with water. To a certain extent, any material can have porous structure and absorb water. From a fundamental point of view of this phenomenon, the structural properties of porous bodies are described below.

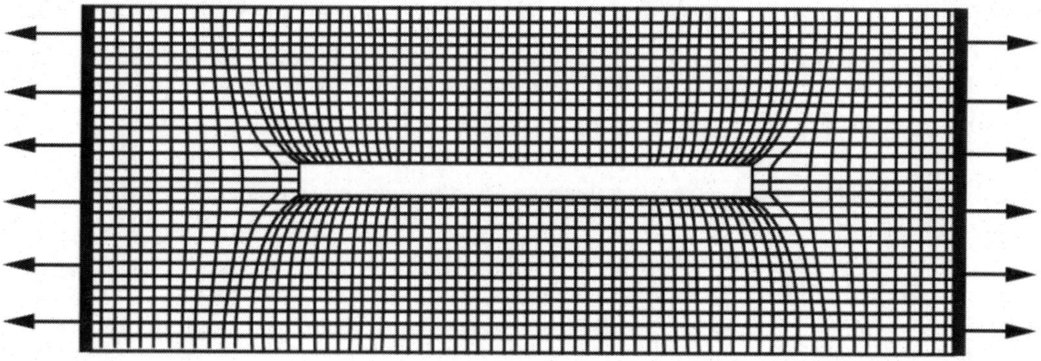

Figure 25. Deformable sheet with embedded bacterial cellulose: the lines show deformation of the sheet when stretched.

A porous environment is considered a solid, containing pores. A geometrical definition of pore concept is hard to determine. Generally, pores are considered hollow spaces distributed inside the solid that can connect or not between each other, being close or open between faces. The connecting portion of the pore is commonly referred to as effective pore space or simply effective porosity. This definition is largely used in filtering technology where porous bodies can have a high total porosity related to a low effective porosity. Total porosity is always superior to the effective one. In general, porosity is defined as the ratio between the pore's volume and form volume:

$$\Phi = V_{pore}/V_f$$

Along with volume porosity, there is the concept of surface porosity, defined as the ratio between the pore's ending area existing onto a sectioned surface of the porous form and the area of the sectioned surface.

The simplest model of the porous body is a system of short radius and equal-form particles (monodisperse system). Any polydisperse system can be fragmented into a monodisperse system by real or hypothetical metric analysis.

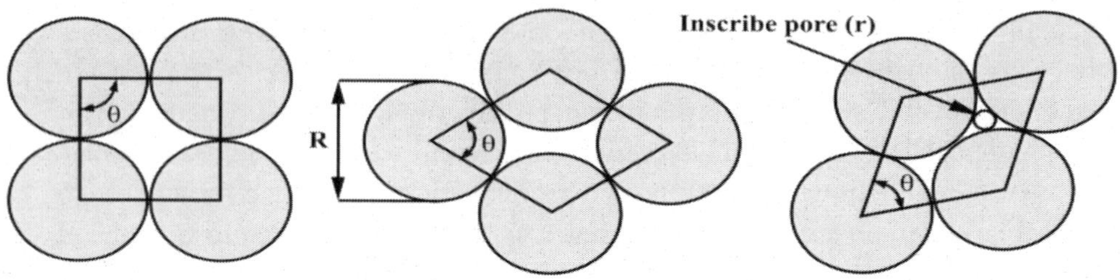

Figure 26. Solid particles layout in forming a cell inside the porous layer.

Figure 26 exhibits that spherical particles can be arranged in different ways. The limits are a rare layout (a, cubic) and a dense layout (c, hexagonal). Between these limits there are all real layouts. In the ideal case, if the particles do not form hydrogen bonds with water, water holding capacity will be directly proportional with the porosity volume [181].

On the other hand, the mechanical properties of composites are generally described below where the pore's volume can be substituted by the matrix volume.

$$P_c = (P_f\, V_f) / (P_m\, V_m)$$

Where:

P_c – Mechanical properties of composite
P_f – Mechanical properties of fiber
V_f – Volume fraction of fiber
P_m – Mechanical properties of matrix
V_m – Volume fraction of matrix

A fundamental parameter of the composite's micromechanics is the representative volume element, for a composite with unidirectional fibers, as sketched in Figure 27.

This volume element represents the smallest portion within the composite material, where stresses and strains are considered macroscopically uniform. Microscopically, these stresses and strains are non-uniform because composite material is not homogeneous. For a 3D orientation of the bacterial cellulose, the microscopic representation is more complicated. For simplification, it will be easier to consider two directions and exact an empirical coefficient for the third direction. In this case, if the wiring geometry is neglected, then two reference dimensions of the volume element are the distances between the bacterial cellulose, one of each direction, and the third dimension is imposed by the number of bacterial cellulose layers in the direction of the material thickness.

The micromechanical analysis of a composite is based on the following simplified hypothesis:

- Lamina is considered macroscopically homogenous and orthotropic, linearly elastic and without initial internal stresses.
- Fibers are considered homogeneous and isotropic, linearly elastic, having a regular position and being perfectly aligned.
- Matrix is considered homogeneous and isotropic, linearly elastic and having a perfect adherence to the fibers.

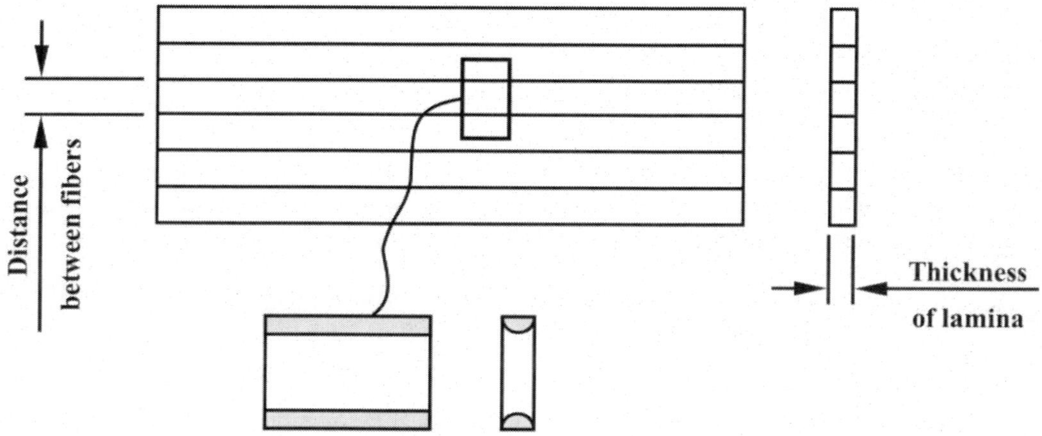

Figure 27. Representative volume elements in composite.

Regarding material porosity, it is acknowledged that hollow spaces inside a porous material can be interior (closed) pores, exterior (open at both ends) pores and inter-granular pores or unoccupied spaces. Monofilament fabrics and granular layers of non-porous particles generally have inter-granular spaces. In operations involving flows through porous structures (filtering, capillary pumping, etc.), the porosity corresponding to inter-granular spaces and sometimes to open pores presents interest. Pore size distribution is also important, though pore "size" or "diameter" is a simplified notion, because pores are far from having cylindrical form. Most commonly, they have irregular forms with contractions and ramifications. This is why a porous material is preferably characterized by the notion of permeability. But there are cases, especially regarding membranes, which are considered as important as the size of the pores.

This theory stands only in the case when cellulose does not form hydrogen bonds with water or when mechanical properties of the material are evaluated for water free environments. However, one of the issues is due to the use of cellulose based composite. Hydrogen bonding and crosslinking of cellulose should be discussed [182-184].

From a fundamental point of view, cellulose has no taste, is odorless, hydrophilic, insoluble in water and most organic solvents, chiral and it is also biodegradable. It can be broken down chemically into its glucose units by treating it with concentrated acids at high temperatures. In pure form, cellulose always absorbs water, which has a plasticizer effect upon it. It is acknowledged that fabrics of cellulosic fibers exhibit good dimensional stability in the dry state, but can shrink and/or wrinkle when wet. This occurs because, in dry state, the cellulose chains are held together by extensive networks of hydrogen bonds between the hydroxyl groups of adjacent chains in its structure. In other words, the hydrogen bonds form a cross-linked structure. If a stress, such as twisting or folding, is applied to the dry fabric, the hydrogen bond cross-links tend to hold the chains in position and cause the fabric to return to their original position when the deforming stress is removed. However, when the fabric is brought into contact with moisture, water molecules can participate in the hydrogen bonding and penetrate between the cellulose chains, effectively breaking up the cross-linked structure. The water molecules act as a plasticizer for cellulose and the chains may move relative to each other. If the fabric becomes wrinkled in the moist state, the chains move to relieve the strain and there is no effective force to return the fabric to its original shape when the stress is removed. Thus, for cellulosic fabrics to exhibit durable press (also termed permanent press or wrinkle resistant) characteristics, it is necessary to form cross-links, which are not easily broken by water. This is usually done with formaldehyde or formaldehyde derivatives, such as urea-formaldehyde resins.

Several research groups have mentioned that, in case of bacterial cellulose, one of the major problems with producing a composite out of this material is the presence of hydroxyl and other polar groups within its structure [185-187]. This makes it highly hydrophilic and therefore it doesn't bond optimally with commonly used resins like epoxy or polyester as they cannot wet the fibers sufficiently. Another problem is the strong crystalline content of cellulose, which inhibits the penetration of resin.

To overcome these issues, some researchers have modified the cellulose by adding hydrophobic elements to the molecular structure and thereby improving the adhesion to hydrophobic resins. One effective route employed silane treatment [188-191]. It reacts with the OH groups of the cellulose and gives it a hydrophobic character [192, 193]. Composites that were made with polyester and epoxy clearly showed an improvement in bonding between the fibers and resin.

When flexural tests were performed the results revealed an increase in flexural strength of up to 40%. However, the author did not report studies on the composite's water sorption behavior. It is important to note that several methods of treating natural fibers have been extensively employed, aiming to improve adhesion characteristics. For example, alkali treatment can remove the hydroxyl group and replaces it with a Na^+ ion, which makes the surface rougher and therefore improves the mechanical interlocking of the cellulose with the resin, but also depolymerizes the native cellulose [194-197]. Graft copolymerization was used to graft vinyl monomers onto natural fibers that are compatible with several resin systems [198-201]. Etherification was done by cyanoethylation and acetylation was used to produce esterified natural fibers, which both result in plastic behavior. Isocyanate groups were used to react with the hydroxyl groups in order to form strong covalent bonds and create better compatibility with binder resins. Treatment with polypropylene is believed to increase the surface energy of the fibers and therefore further increase the wettability and interfacial adhesion.

Very good results were also obtained when cellulose was used as matrix. In 2004 Nishino et al., [202] published the Paper "All-Cellulose Composites," in which they characterize a new type of composite having cellulose as a matrix and cellulose fibers as reinforcement fibers. In their studies, they used refined ramie as cellulose fibers and for the matrix they used pretreated craft pulp from coniferous trees solved in 8% LiCl. The new composite revealed a tensile strength of up to 480 MPa, this strength being the highest tensile strength ever reported on any cellulose based composite. However, the author did not report studies on the composite's water absorption capability.

3.2.1. Example of Cellulose Based Composite

Cellulose based composite was theoretically defined as two phase materials (the one was cellulose, the other one was commonly considered on polymer matrix). The importance of cellulose based composite not only provided significant efforts in terms of biodegradability and favorable CO_2 balance growth with the awareness of consumers for sustainability in the use of materials but also offered the value-added concept of agricultural product and waste. Compared to conventional composite, cellulose based composite was highly considered for any challenging idea in both engineering and medicine research communities.

Cellulose is one of the most abundant naturally occurring biopolymers; commonly founded in the cell walls of plants and certain bacterial species. Before using cellulose in applications with higher efficiency, the modification of both physical and chemical properties of cellulose by means of any pre- and post-treatment was mandatorily required. The incorporation of cellulose into polymer matrix can effectively lead to enhanced mechanical properties as well as thermal resistance if cellulose based composites are applied in applications where heat generation are involved. Interestingly, it was commonly known that outstanding properties of cellulose were due to high thermal resistance. A small amount of cellulose can be effectively inserted into polymer matrix, yielding not only excellent thermal resistance in polymer but also more stiffness from cellulose can offer excellence in flexibility. Up to the present time, cellulose based composite has been extensively researched. It was very well known that cellulose has emerged in industrial commercialization in composites as reinforcement with both petrochemical-based polymer and bio-based polymer matrices. The example of cellulose based polymer matrix composite and its comprehensive potential application in research was discussed.

In the framework of environmentally friendly processes and products, polylactic acid was considered the most effective biopolymer for bio-based composite, owing to the concept of renewability, biodegradability as well as biocompatibility [203-205]. From the viewpoint of preparation route, polylactic acid was a biodegradable thermoplastic polyester manufactured by biotechnological processes from renewable resources. Corn, common source of biomasses, was the most effective source for providing high purity polylactic acid. The design of polylactic acid based composite was strongly encouraged to be developed for green scenarios. Recently, producers of polylactic acid formally announced a production capacity of 140,000 ton/year for the commodity market. It can be consequently considered that polylactic acid gained more and more interest as a valuable bio-resource for any alternative application.

From the viewpoint of polylactic acid research community, it was well-known that polylactic acid exhibited low molecular weight and relatively low thermal resistance [206-208]. In addition, in the polylactic acid synthetic route, it commonly required high temperature in the polycondensation process, continuous water removal, as well as long reaction time. This consequently reduced the efficiency of polylactic acid in industrial commercialization. By the way, relevant engineering properties of polylactic acid can be significantly improved by modifying polylactic acid based composite form. The reinforcement with cellulose was strongly encouraged. Since the properties of cellulose were evident in terms of mechanical and thermal resistance including high stiffness, low density, unique morphology, and high aspect ratio as well as high melting temperature. The incorporation of cellulose into polylactic acid can not only adjust engineering properties of polylactic acid but also open wider vision where the use of polylactic acid can be extensively addressed.

Although significant properties can be effectively enhanced by insertion of cellulose into polylactic acid, the major challenge in the incorporation of cellulose into polylactic acid matrix has still concerned its dispersability. It was commonly known that the nature surface of cellulose exhibited hydrophilic properties while polylactic acid surface exhibited hydrophobic properties. This subsequently made cellulose become aggregated in polylactic acid matrix. In addition, irreversible agglomeration of cellulose also occurred during the drying procedure because of strong hydrogen bonds established through the large number of hydroxyl groups present on the surface. The design on composite between cellulose and polylactic is very interesting and its application can be versatile. Cellulose and polylactic acid composite have been enrolled in many application areas including automotive parts and infrastructure.

On the other hand, PHB has received much attention on bio-based research in recent years with a large number of publications ranging in topic from biosynthesis, microstructure, thermal and mechanical properties as well as biodegradation [209-211]. Numerous research works have been driven by the availability of PHB from renewable resources and the similarity of PHB relevant properties to those of conventional plastics. The design of PHB and cellulose composite has been evident up to the present time. In the past, cellulose and PHB composites were prepared by using hydroxyvalerate contents as raw materials. The amounts of raw materials, biodegradability and the influence of manufacturing condition on engineering properties of composite were evaluated. It was important to note that this composite exhibited low impact strength if injection molding was used to fabricate. The tensile strength significantly decreased in the initial stage of degradation. Due to this reason, many approaches have been developed on the interface of PHB based cellulose composite. Interfacial properties improvement can be enhanced by addition of 4,40-thiodiphenol at various concentrations. This additive can effectively form hydrogen bonds with many functional groups at the surface of cellulose, leading more adherent features of cellulose in PHB composite. On the other hand, prior to composite preparation between cellulose and PHB, chemically modified fiber by PEG grafting on cellulose was investigated. The better adhesion can be subsequently observed if this chemically modified cellulose was fabricated with PHB by compression molding. More and more research on PHB can be extensively developed and applied in many forms of cellulose. The engineering properties including mechanical, thermal and thermo-mechanical properties were consequently evaluated. Up to the present time, the application of PHB and cellulose based composite was enrolled in medical and pharmaceutical areas as well as sensor materials.

In recent years, progress in polyurethane based composite was extensively reviewed. In the past, it was increasingly used in a variety of applications due to comfort, cost benefits, energy savings, and potential environmental soundness. The various types of polyurethane products were generally prepared in the form of flakes, powders and pellets. In industry, it can be found that the application of polyurethane and polyurethane-based composite can be versatile [212-215]. It can be prepared as foam-like structures for automotive interiors, foot wear, and medical devices. From a structural point of view, polyurethane exhibited excellent mechanical properties, high thermal resistance as well as high chemical resistance. The role of polyurethane was subsequently employed in many engineering sectors. In addition, for cellulose composite community, polyurethane was developed as many ways of composite. It can reduce water absorption ability to cellulose and provide high flexibility. Recently, it was found that the use of polyurethane as matrix with cellulose was related to electronic substrate. PU can lead to enhanced mechanical properties and reduce the water absorption of cellulose. This composite exhibited remarkable application as an effective organic light emitting diode substrate with additional features of flexibility and transparency. On other hand, polyurethane can be effectively used in membrane separation technology. Cellulose-based polyurethane can increase the stability in any pH media. Permeability and swelling characteristic of membrane can be controlled to have a long life-time for separation technology.

Here, the detail presented above is the fundamental theory of cellulose and cellulose based nanocomposite materials. The highlight of this research is to develop cellulose based composite as a flexible substrate for organic light emitting diodes (OLEDs).

3.3. Modification of Cellulose for Flexible Substrate

3.3.1. Cellulose Modification for Composite

In recent years, numerous efforts have been extensively developed on the properties of cellulose and its investigation for many applications. The role of modified cellulose was versatile in many industrial sectors. The example was due to infrastructure. It was important to note that many successful projects on the use of cellulose for infrastructure were due to the development of wood based plastic composite. Before using cement and concrete for infrastructure, wood based material was versatile and it played an important role in infrastructure-based materials. However, due to the shortage and the concern over the environmental impact of the use of wood from forests, many alternative materials have been developed. Cement and concrete may gain the interest due to the availability of raw materials and cost effectiveness. Moreover, the use of wood based material is still being pushed due to Green Technology. Many processes and materials that involved on the use of renewable and sustainable concepts were encouraged for use in many engineering sectors. Moreover, the development of cellulose has been extensively developed for automotive parts. Due to its strength and stiffness at elevated temperatures, cellulose gained much interest for development in automotive parts. The use of cellulose can overcome the limitations of previous materials in automotive parts. In the past, it was remarkable that material for automotive parts strongly used metal-based materials for the design as structural materials. This was probably due to its strength and high resistance to environmental impact. In addition, the reason for the rising price of crude oil has exponentially gained impact due to environmental concerns, it was subsequently noted that the insufficient quality and quantity of crude oil were low and many strategic methods have been proposed on the use of automobiles with less utilization of crude oil. One of the effective strategic routes considered the weight of automotive parts. It was important to note that although the use of metal-based material gained much effectiveness due to high strength, thermal resistance and any high resistance to environmental impact, the use of metal-based materials for automotive parts is limited due to weight. Consequently, the use of metal-based materials for automotive parts should be reconsidered because of the cost of fuel, which was commonly referred to the consumption and production of crude oil. As mentioned above, due to the shortage of fuel from crude oil and since its price has been exponentially raised, cellulose has gained much interest as replacements for metal-based materials. In the previous chapter, cellulose exhibited many outstanding properties such as high stiffness, high thermal resistance, low co-efficient of thermal expansion and high chemical resistance. Therefore, due to these properties of cellulose, modifications on cellulose can be rendered for many prospective applications. Regarding automotive parts, cellulose has been physically and chemically modified to replace the use of metal based materials. Moreover, the role of cellulose has also gained much interest in medical and pharmaceutical based research. Due to the process of cellulose extraction, which was commonly involved in water-based suspension, cellulose easily adsorbed water and humidity. Although some applications preferably avoided moisture absorption, the high water absorption ability of cellulose encouraged strong interest in the medical and pharmaceutical area. It gained many efforts due to absorption and desorption of water in order to control the swelling behavior. The design of cellulose for drug delivery systems should be investigated regarding the swelling behavior of cellulose and also the effect of pH, temperature and enzymes in order to evaluate the ability of cellulose and drug model for its control and release mechanism. Moreover, this concept was also extended to food-based materials; the design of active packaging for food technology should consider the high effectiveness of packaging, which exhibited strong ability on moisture and vegetable oil sorption.

Up to the present time, the role of cellulose has gained interest for electronic devices. Examples of electronic devices are versatile, such as energy storage devices, fuel cells, diodes and transistors [59, 140, 216-219]. In the meantime, numerous efforts have been developed and have improved the quality of cellulose for use in many areas of electronic devices. Among many types of electronic device, cellulose modification gained interest as being substrate in display technology. In the past, the role of display technology has been utilized in many applications. Examples include light source media and television. The use of flat glass sheet was enrolled as substrate. Up to the present time, the trend of electronic device was preferably enrolled on flexible feature of substrate. This enabled the push to development of cellulose-based materials. The evaluation on feasibility of flexible substrate from cellulose has gained strong interests for development.

However, before employing cellulose-based materials in flexible display technology, there have been numerous concerns on the use of cellulose and its related modification techniques. Figure 30 exhibits the morphological properties of cellulose. It can be found that structure of cellulose was a network of fibril and it resulted in high strength and stiffness.

It was mandatory to note that the use of cellulose in electronics was very limited due to the water absorption efficiency of cellulose. Moreover, it was notable that due to the high water absorption of cellulose, applications for cellulose were often limited to interior and non-structural applications and it affected lower mechanical properties and had poor moisture resistance. To employ cellulose with higher efficiency, significant improvement using innovative surface treatments, resins, additives, and coatings had to be considered. Other challenges facing cellulose involved on fire resistance, cellulose/matrix adhesion, durability, variability in quality, and manufacturing difficulties.

From its nature, cellulose was considered for its hydrophilic properties. This was the main drawback of cellulose, as this caused it to have high water absorption and to be incompatible with hydrophobic polymer matrices. Compared to glass fiber, it presented high moisture resistance. The design of glass fiber based composite as building materials has failed in wet conditions through surface roughening by swelling of cellulose or delamination. From the fundamental point of view, the design of cellulose-based composite should carefully consider the effect of water absorption. It performed particularly poorly when immersed in water.

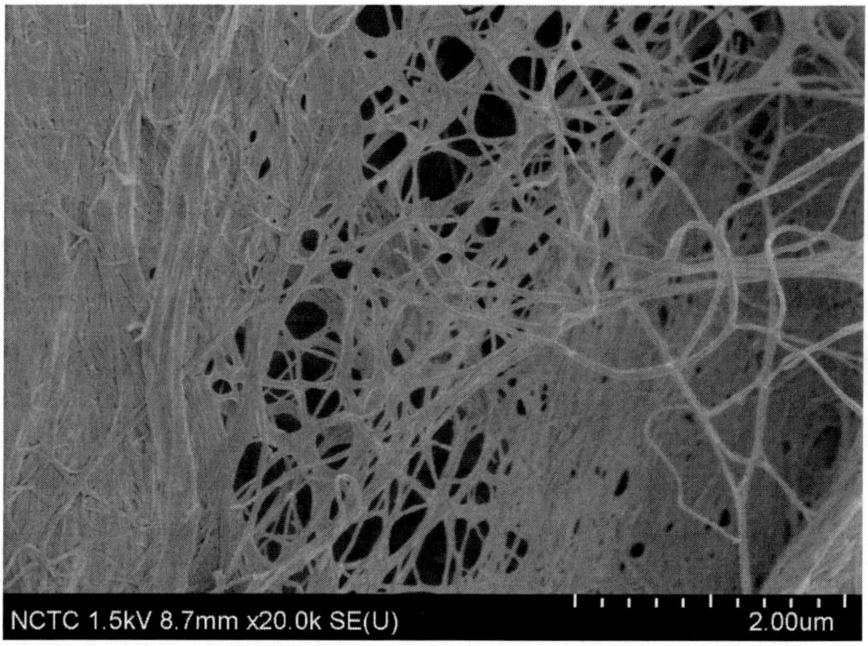

Figure 28. Morphological properties of cellulose.

Before using cellulose in electronic based substrate in display technology, numerous efforts extensively evaluated the water absorption ability of cellulose. The most common way to reduce the moisture absorption capability of cellulose was through the process of alkalization. The process of alkali treatments such as KOH and NaOH can effectively reduce the hydrogen bonding capacity of the cellulose, eliminating open hydroxyl groups that tend to bond with water molecules. Moreover, the process of alkalization can dissolve hemicellulose. Removal of hemicellulose, which was considered the most hydrophilic part of cellulose, reduced the ability of the cellulose to absorb moisture. Moreover, due to the use of the alkalization process, the amount of alkali and time should be controlled. It may affect the properties of cellulose so the amount of alkali should be removed. It was necessary to adjust the neutralization by washing with a low-concentration of acid and distilled water. If the cellulose was not carefully washed after alkalization, the alkali continued to degrade the fibers long after exposure, eventually leading to significant fiber swelling and breakdown. The use of alkalization should also consider its negative effect. The example of negative aspects of the alkali treatment process includes high pH values, high surfactant content, polluted wastewater, and the degradation of cellulose. Alternatively, the other process used to decrease the water absorption ability of cellulose involves pectin-dissolving enzymes. It can also be more effective in lowering moisture absorption; the enzyme-treated fibers produced less polluted wastewater. Moreover, use of the steam treatment process was one of the most effective ways for water absorption reduction. It exhibited great promise on moisture absorption ability. Moreover, steam treatment presented numerous advantages such as better dimensional and temperature stability, better resistance to fungal attacks, and generally improved mechanical properties. Up to the present time, significant properties of cellulose can be effectively enhanced using the steam treatment process.

After chemical modifications of cellulose using the alkalization technique, cellulose-based composite was developed. Up to the present time, from the fundamental point of view, the design of cellulose has been exponentially developed. It is remarkable to note that the effective design was considered a sandwich-form of cellulose based composite. Cellulose was prepared as a thin flat sheet and it was encapsulated on both sides. In order to process cellulose as a flat sheet, the cellulose preparation was similar to the papermaking process. Cellulose based water suspension was prepared in the appropriate volume fraction. It can therefore be rendered on weight and thickness of cellulose. In this feature, the role of encapsulated material should focus on polymers. Since we expect flexibility in composite, matching between polymer and cellulose should be controlled. The objective of polymeric-based material on both sides of the cellulose was considered for water absorption ability and to retain the flexibility of cellulose. Figure 31 exhibits the sandwich-like structure of cellulose. In general, cellulose was prepared as a sheet and it was encapsulated on both sides for less water absorption ability.

Up to the present time, next to the alkalization process, cellulose was extensively developed in order to reduce the water absorption ability. In order to solve this issue, hydrophobic properties of polymers should be developed. It can be encapsulated on both sides of cellulose, rendering significant effects in terms of water absorption ability. An example of polymeric-based materials being used as encapsulation involved polyurethane, polyether and polyethylene oxide.

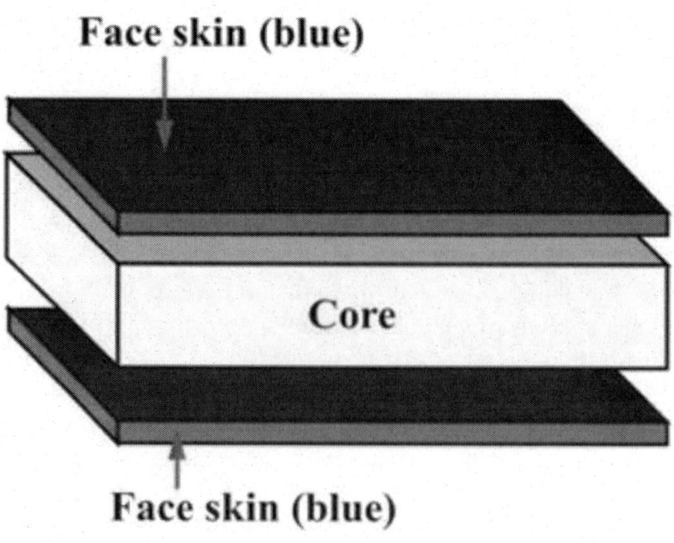

Figure 29. Sandwich-like structure of cellulose.

Moreover, in order to gain the most effective polymeric-based material, the criteria on polymeric-based material should consider its adhesion and wetting ability with cellulose.

From a fundamental point of view, the development of cellulose and polymeric-based material, called bio-based composite, has exponentially received much attention due to the shortage of petrochemical based polymeric materials. Most 'semi-green' or 'green' composites have been extensively designed. From a fundamental point of view, cellulose based composite exhibited excellence of tensile strengths and stiffness in the range of 100–200 MPa and 1– 4 GPa. Moreover, there have been numerous methods for cellulose based composite modification. The design can be handling by structural configurations that negate these disadvantages. The example of design was considered a hybridization process, which exhibited much potential to improve mechanical properties.

In order to gain the most effectiveness of the design of cellulose based polymeric composite, adhesion and wetting ability was one of the most important key factors. Many researchers have been extensively developing the chemical and physical treatments to modify the surface characteristics of cellulose, which can improve compatibility and adhesion. From the fundamental point of view, three factors that affect the mechanical properties of cellulose based polymeric composite. In general, the mechanism between cellulose sheets and polymeric-based composite should consider mechanical interlocking, molecular attraction forces and the chemical bonds between the cellulose and polymeric matrix. However, due to the nature of cellulose chain, the reaction between hydroxyl groups in the polymeric matrix would bond with the hydroxyl groups, subsequently creating hydrogen bonds. However, one of the utmost concerns was decreased bond strength caused by water absorption. It was remarkable to note that the bond strength between cellulose and polymeric matrix was significantly lowered by the existence of moisture during the polymerization step. Moreover, the other criteria involved removal of water. When it was evaporated, it existed the void in the structure of composite, suggesting the process of cellulose drying should be carefully controlled. It resulted in bond strength and the existence of voids. Figure 30 exhibits the morphological properties of cellulose sheet and polymeric matrix. It can be noted in between the cellulose and polymeric matrix, called the interface region. It was important to note that if there is excellent compatibility, there are no detrimental effects on the interface.

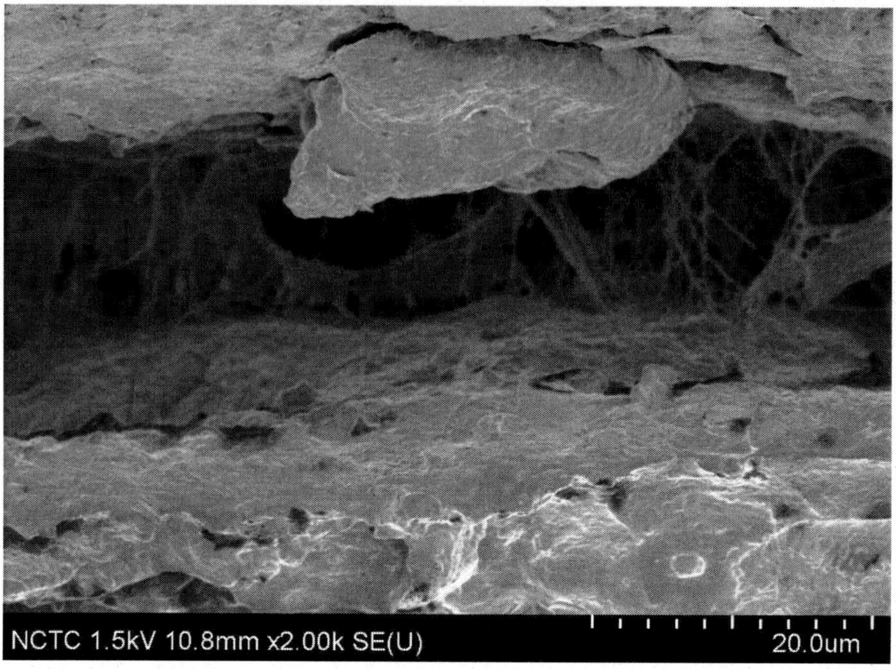

Figure 30. Morphological properties of cellulose and polymeric-based composite.

Up to the present time, from the viewpoint of flexible electronic substrate, the concern over moisture absorption ability of cellulose is still being discussed. One effective route to overcome this concern related to the use of a coupling agent. In the past, it was remarkable to note that significant enhancement of the compatibility between the cellulose sheet and the polymeric resin was due to the effect of alkalization. Alkalization was introduced as a third material and it remained in the composite in order to provide the good support between the cellulose and hydrophobic polymer. In order to gain more efficiency of the cellulose based composite, the use of a coupling agent was required. It improved the bond between materials, eliminating weak boundary layers, producing tough, deformable layers, by developing a highly cross-linked interphase region with an intermediate modulus, improving the wettability and critical surface tension factor, forming covalent bonds with both materials, and altering surface acidity.

Up to the present time, the role of coupling agent was very necessary and its use was recommended in order to obtain the excellent performance of composite. An example of a coupling agent, currently employed, is silane treatment. It can form chemical bonding between cellulose and the hydrophobic polymer. Silane was successfully developed and it was versatile in many molecular structures. The example silane involves trialkoxysilanes. In general, it was employed as the coupling agent for the inorganic reinforcement and hydrophobic polymer matrix. However, up to the present time, the role of this silane coupling agent was for cellulose based hydrophobic polymer. With proper surface or bulking treatment with silanes, hydroscopic cellulose can be converted into a hydrophobic reinforcement. Surface treatments decrease the water sorption rate, while bulk treatments additionally reduce the water sorption capacity. It was remarkable to note that the role of coupling agent was different from alkalization. It exhibited less effect on properties of cellulose-based hydrophobic composite. However, with the use of a coupling agent, significant enhancement on higher interfacial shear strength in cellulose based hydrophobic can be obtained. The other example of coupling agent for cellulose and hydrophobic polymer involves stearic acid. Stearic acid can cause the structure of cellulose to be hydrophobic and it can improve the dispersion state and wetting ability. However, there have been numerous coupling agents. The use of isocyanate can be chemically linked to the cellulose through strong covalent bonds. Triazine can also form the covalent bonds with cellulose and can reduce moisture absorption. Moreover, the use of acetylation was considered to have the most effective reaction. It involved an esterification reaction to introduce plasticization, and has been shown to improve dimensional stability and environmental degradation. Maleic anhydride grafted polypropylene was considered the common compatibilizer for cellulose and hydrophobic polymer matrix. It can improve the wettability and adhesion.

Moreover, for reduction of the water absorption ability of cellulose, the feasibility of an additive was therefore investigated. Inserting a small amount of inorganic material or ceramic particles may reduce the water absorption ability of cellulose. The example of additive involves calcium chloride and sodium carbonate particles, which can be inserted in the porous structure of cellulose sheet. Ceramic and inorganic particles on the interface between cellulose and hydrophobic polymer matrix may subsequently reduce mechanical properties.

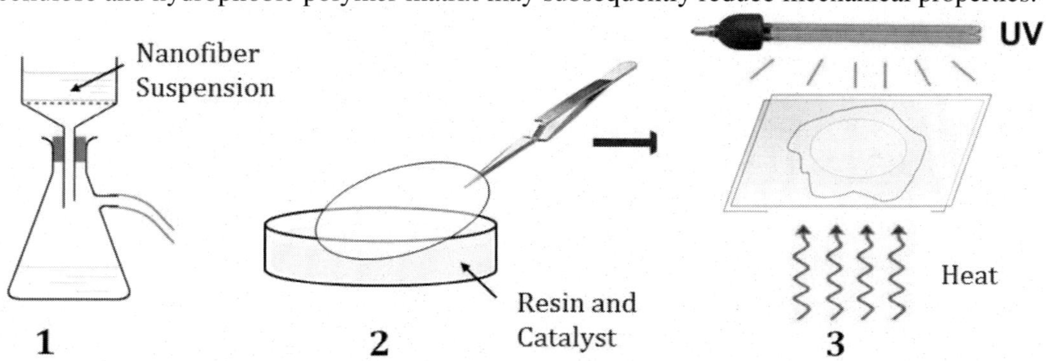

Figure 31. Cellulose and polymer matrix preparation by UV irradiation.

One of the biggest concerns regarding the use of cellulose-based composite was durability. The objective of modification of cellulose as composite based material was a sustainable and renewable concept. Many researchers have investigated its feasibility, not only as flexible substrate in display technology, but also in other applications such as infrastructure, automotive parts and medical devices. There are concerns regarding durability of cellulose-based composite and its long-term behavior when exposed to different environments. Different environments involve hydrothermal aging and loading as well as prediction of lifetimes. Fatigue work is a big concern for cellulose-based composite. It was considered the main structural component before designing cellulose-based composite structure. The difference of design involved the amount of cellulose and hydrophobic polymer matrix, orientation and processing techniques. Furthermore, the preparation of cellulose involved water-based suspension. The durability of cellulose-based composite due to moisture absorption and weathering has been discussed. Water absorption can render the growth of fungi and bacteria in cellulose based composite. It can therefore result in swelling behavior and bio-degradation of cellulose-based composite. Performance in any application may be reduced. Therefore, numerous methods have been developed on the use of cellulose-based composite, far away from moisture absorption. Moreover, there has been concern over UV irradiation to cellulose based composite. UV can cause the color fading. Cellulose based polymeric composite should not be used in activities related to UV exposure. UV exposure can lead to cracking, bulging, fibrillation, and black spots. It can therefore result in reduction on tensile strength. Moreover, UV exposure can induce color change, weight loss and surface roughening. It is suggested that before using cellulose-based composite, a UV protection technique should be considered to prevent surface deterioration.

The other concern regarding the use of cellulose-based composite involves variability. Cellulose is one of the three main constituents of a plant. It is the most abundant polymer in the world. It was estimated that the amount of cellulose on earth is very high. The cellulose content in plants varies for different species. The properties of the cellulose differ from plant to plant depending upon the age, environment and soil conditions. The properties of cellulose are altered during certain industrial processes such as alkalization and coupling agent treatment. Cellulose was produced extensively as a co-product in the bioethanol and paper pulp industries. During ethanol and paper pulp production, removing lignin and any impurities from the system was one way to obtain a good quality and quantity of cellulose. Moreover, not only do environmental conditions affect plant growth, but processing steps also provided many significant properties for cellulose. Examples of environmental conditions causing variations of cellulose involves crop variety, seed density, soil quality, fertilization, field location, fiber location on the plant, climate, weather conditions, and harvest timing, which can all impact the raw material. On the other hand, additionally, variation in the cross-sectional area of the fibers, extraction processing methods, damage incurred in handling and processing, and differences in drying processes can lead to further variation in the end-use product. It was remarkable to note that for industrial commercialization, cellulose quality assurance protocols were necessary to ensure that the cellulose product was uniform and so that dimensional variability would not lead to drastically different mechanical properties between production batches. Moreover, the variation of cellulose and its processing steps should be considered due to the price of raw materials. An example is the comparable price of cellulose derived from flax, which was very expensive due to climate variations.

To use cellulose with higher efficiency, the manufacturing process of cellulose-based polymeric composite should be considered as the most important step. From a fundamental point of view, the manufacturing and processing of cellulose-based composite was considered, based on the ability of polymerization of polymer matrix. In general, the temperature of processing was set to 200°C, depending on the glass transition temperature of polymer. Prolonging polymerization time may be required while the consideration of mechanical properties should be carefully controlled. Up to the present time, many polymers have been investigated and evaluated as to their use as a composite. The simplest method for cellulose-based composite preparation involves hand lay-up process.

The advantage of this process did not intrinsically involve the application of heat. Moreover, the other manufacturing process considered the compression technique. It was the preferable technique for mass production of cellulose-based composite. Moreover, the use of resin transfer molding and vacuum molding may be employed for cellulose-based composite with higher fiber volume fractions. Vacuum-assisted resin transfer molding was able to polymerize in large-volume of cellulose and polymer based composite at room temperature. Moreover, the consideration on pultrusion technique is still interesting. It was employed on cellulose based composite. The exploration for pultrusion technique for cellulose can render high potential for use with cellulose and bio-based polymer.

Up to the present time, the process of cellulose based composite preparation was versatile, depending on the polymerization technique of polymer, reaction time and the design of its prospective application. From the viewpoint of flexible substrate in display technology, cellulose was designed as a flat sheet. In the past, the quality of cellulose was adjusted from a suspension-based process. Controlling its size and uniformity helped obtain excellent performance in application. In this case, after physical and chemical treatment of cellulose, the dimension of cellulose was prepared into nano-scale level. The mechanical grinding method was employed in order to lower the cellulose network from micro-fiber to nano-fiber level. The dimension of the nano-scale cellulose was very low and it was good enough to be fabricated in optic-based substrate in display technology. Figure 31 presents the cellulose sheet preparation process. It involved cellulose suspension. Water was removed and it was subsequently immersed in liquid polymer. The polymerization was set by the operation of UV irradiation technique. Cellulose and polymer matrix as a composite was therefore prepared.

From the fundamental point of view, the UV polymerization or UV curing was versatile and it was exhibited in many applications [220-224]. In general, the process of UV polymerization was used to improve the quality of the finished product. It gained much interest due to its productivity. It was considered the primary technique because of substantial savings in energy, processing time and cost of labor. In the UV polymerization process, UV light interacted with specially formulated un-cured liquid polymer to polymerize coatings faster and more economically than traditional methods. In reality, the role of UV polymerization was versatile. It can be employed in printing technology, active packaging, electronic devices and wood-based products. Up to the present time, UV polymerization was used to position shape coating and ink technology. The product and process of UV polymerization provided many advantages such as environmental benefit, unique physical properties, and production efficiency.

The low levels of volatility provided environmental benefits.. To be sustainable, volatility should be controlled. Products and processes that prevent the use of hazardous substances should be avoided due to environmental regulations on limiting emissions and controlling air quality. An example of the unique physical properties of UV polymerization is called enhancement technology. The UV polymerization technique was considered the most effective method for designing the product in order to be more attractive and durable. It was preferable that the finished product should exhibit exceptional strain abrasion, and the environmental strength should be superior. Moreover, the production efficiency of cellulose-based polymer composite has exponentially gained much efficiency. UV polymerization can be installed on an existing line and it can lead to reduction of required factory space. From the fundamental point of view, it required light sources that can direct UV or visible light onto the formulated product. The use of a photo initiator can absorb UV energy from the light source, setting in motion a chemical reaction that quickly, in fractions of a second, converted the liquid formulation into the solid form. Figure 32 exhibits the mechanism of the UV polymerization process.

From the chemical reaction, the bulk of the formulation used monomer and oligomer as a starting material. Compare to bulk polymer, monomer and oligomer were considered low molecular weight materials and they were therefore converted to polymer by the reaction activated by UV polymerization. They can be monofunctional and multifunctional molecules, depending on the number of reactive groups. The use of UV exposure can change a functional monomer to become part of the polymer matrix. Moreover, monomer was considered a functional material. It was diluent in the formulation and viscosity adjustment was needed. However, the use of oligomer was different because of its larger molecular weight and it exhibited the more viscosity compared to monomer. From the fundamental point of view, a functional group of oligomer was employed to determine the final properties of cellulose-based composite. The example of properties was due to flexibility and toughness, etc.

The UV polymerization process considered the effect of photo-initiator [225-228]. It was considered an additive that initiated the polymerization process. The process of polymerization was related to the curing step and the structure of monomer and oligomer, which reacted to form a polymer network. The role of photo-initiator was very important since it helped activate any reactive monomer and oligomer. In reaction, as UV light energy was emitted, it was absorbed by the photo-initiator, causing it to fragment into reactive species. These species can be either free radical or cationic. In general, the majority of systems were based on free radicals, which reacted with the unsaturated compounds in the liquid formulation and subsequently caused them to polymerize.

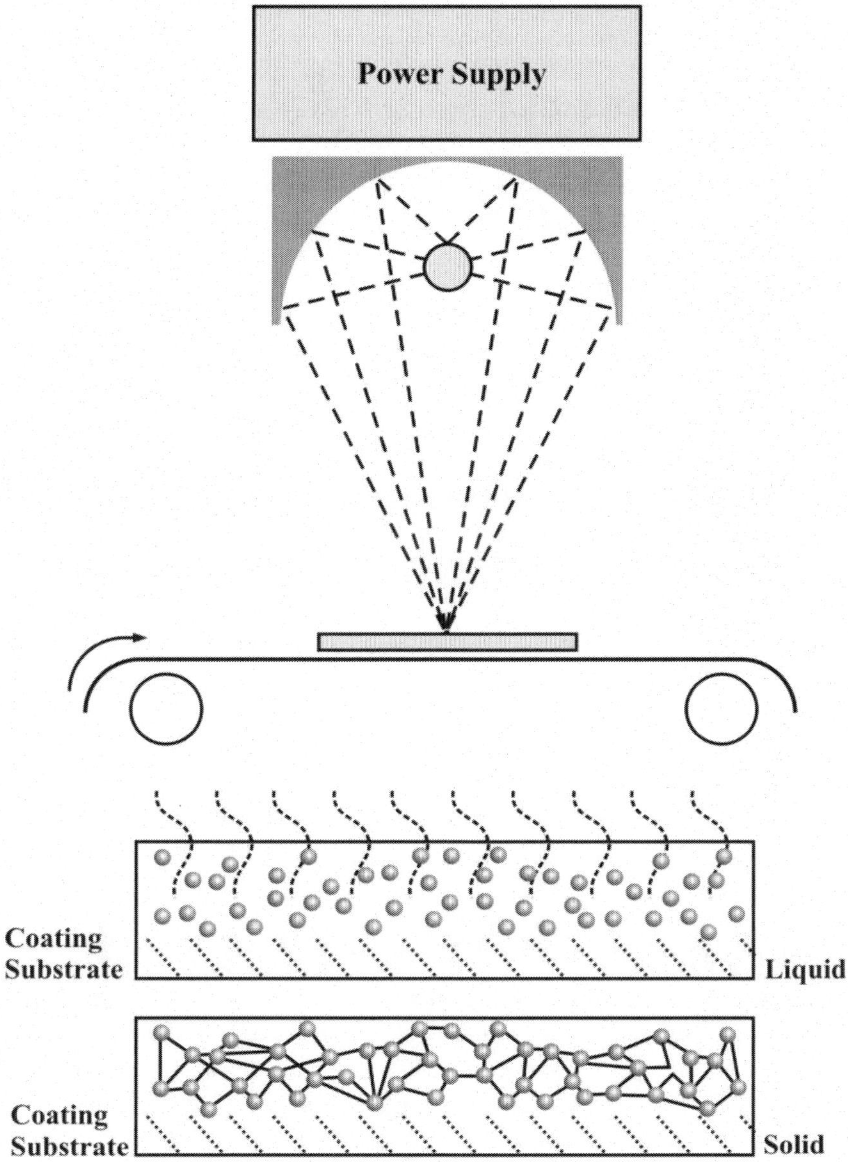

Figure 32. Mechanism of UV polymerization process.

In addition, since the active reaction of UV exposure was related to photo-initiator, the UV light source was also important. From the fundamental of view, UV light source can be categorized into two different sources, arc light and laser light. The arc light involves the medium pressure mercury lamp and the high-pressure xenon lamp. The medium pressure mercury lamp generally used high power. In industrial commercialization, the use of arc light was well known. Therefore, it resulted in many lamp developments, which can be offered in specific wavelengths. Another development involved the excimer lamp. It can take out a large portion of IR radiation to allow the UV exposure of the heat-sensitive substrate. Moreover, in order to gain more efficiency of UV exposure, another approach using a diffused UV lamp system was preferred. However, to use UV exposure with higher efficiency, the emission of the lamp must overlap the absorption spectrum of the photo-initiator.

For the photo-initiator, the use of free radicals was commonly employed in industry. It can be categorized into either hydrogen abstraction types or alpha cleavage types. In general, the hydrogen abstraction types depend on the specific use, especially in three- or four-way photo-initiator mixtures. However, on the other hand, the use of alpha cleavage initiators was more versatile and subsequently the most widely employed. Alpha cleavage initiators generally have higher efficiency due to their generation of free radicals via a unimolecular process. It requires absorbing light in order to generate radicals. On the other hand, the use of a hydrogen abstraction initiator required an extra step after light absorption. It can be noted that the excited state photo-initiator must find a hydrogen-donating source in order to generate the free radicals. In general, from the fundamental point of view, the use of photo-initiator should preliminarily consider the region of wavelength of light. Therefore, the ability of absorption should be focused. In general, the photo-initiator that is commonly used in industry should be focused on the wavelength region of 300-450 nm. Longer wavelength light was required to provide the thicker sample.

In order to evaluate the use of UV exposure, a photo-initiator and a type of functional polymer should be considered. The development of cellulose-based composite therefore employed the UV exposure. It was necessary to understand the prerequisite requirement of the sample. The example related to the thickness of the polymer in the composite and surface properties, and it therefore gained potential for being developed in mass production in industrial commercialization.

3.3.2. Properties of Cellulose Based Composite as a Flexible Substrate in Display Technology

Before using cellulose-based composite as a flexible substrate in display technology, the properties of cellulose-based composite should be discussed. One of the highest concerns regarding properties of cellulose-based composite regards optical properties. Note that the dimension of cellulose was scale-down to nano-range. The typical dimension of cellulose is 10 nm in width and 50 nm in thickness, which is smaller than one-tenth of the visible light wavelength (400–800 nm); therefore, it is free from light scattering [57, 229, 230]. Figure 33 exhibits the cellulose-based composite as a sample for flexible substrate for display technology and only cellulose sheet is provided for comparison.

In Figure 35, the sample involves nano-scale cellulose and its modification as composite. From the fundamental point of view, it is possible to prepare the nano-scale of cellulose from the material that is derived from plants and bacterial species. As mentioned above, for plant-based cellulosic materials, involving a top-down approach, the break-down of fiber was employed and chemical modification was used to purify cellulose particles. However, on the other hand, for cellulose derived from bacteria, it involved a bottom-up approach; the growth of cellulose particles was prepared. The advantage of cellulose derived from bacterial sources involves uniformity in size and shape and fewer impurities. Figure 33 exhibits that cellulose sheet is not optically transparent and that the composite is clearly much more optically transparent. Cellulose sheet can be considered a three-dimensional network structure of nano-sized fibers with air interstices in between. The opacity of the bacterial cellulose sheet can be ascribed to the light diffraction at the interface.

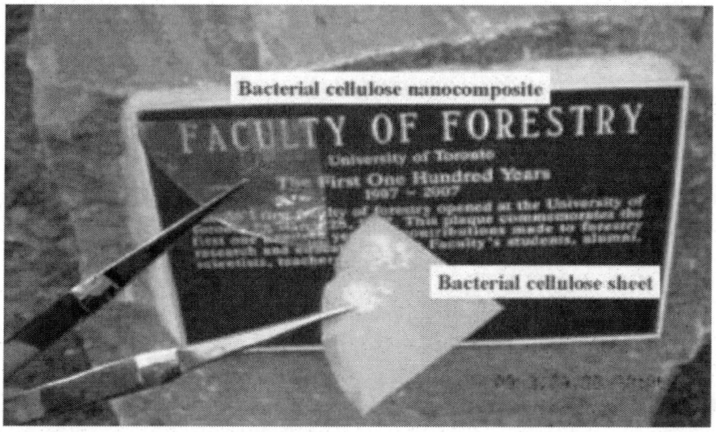

Figure 33. Cellulose based composite as a sample for flexible substrate for display technology [59].

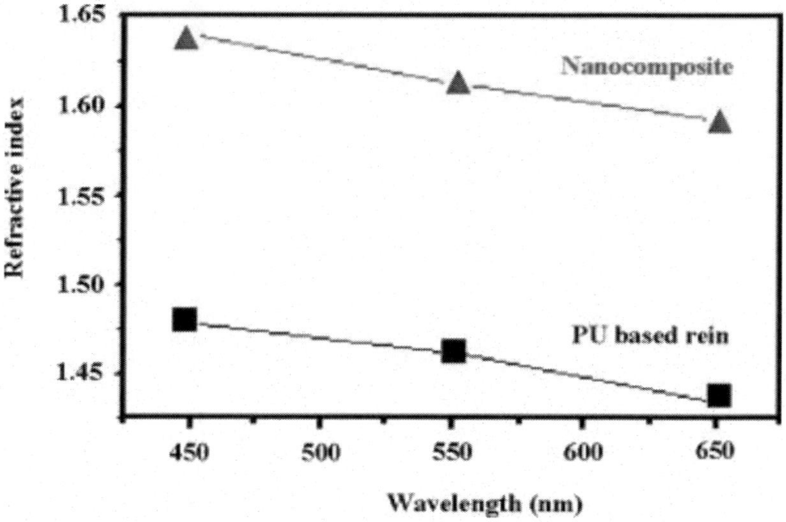

Figure 34. Refractive index of the cellulose based composite [59].

Moreover, not only the transparency but also the refractive index was evaluated since the material was composed of two different parts. One was cellulose while the other was polymer matrix. The refractive index of both materials should be preferably controlled in the similar region. This was probably due to effect of light diffraction. Since the application involved an optical device, it is important to prevent the effect of light scattering. The selection on polymer matrix should preferably consider the refractive index of polymer matrix closer to 1.5. This data was estimated as the refractive index of cellulose in nano-scale region. In general, the refractive index of cellulose-based composite was evaluated as a function of wavelength. For only cellulose sheet, the refractive index was estimated to be 1.5. It was therefore remarkable to note that the important key factor of the design of composite related to polymer matrix. The refractive index of polymer matrix should be approximately 1.5 as compared to cellulose sheet. The example of polymer matrix was focused on polyurethane based resin. Moreover, one of the areas of concern in composite formation involves air interstice. It exists in the area between cellulose and polymer matrix. From the fundamental point of view, cellulose-based composite did not contain air interstices; the interstices were replaced with the polyurethane-based resin of which the refractive index is close to that of the cellulose sheet. The matching refractive index of the polyurethane-based resin and the cellulose means that the light diffraction at the interface between the two components will be restricted and the transparent film can be achieved. The refractive index of the composite and the neat polyurethane-based resin were measured as shown in Figure 34.

Moreover, the application of cellulose-based composite was involved in the optical device. Transparency of the sample was therefore concerned in the design of composite. From the fundamental point of view, an electromagnetic wave can be categorized by the use of wavelength. It can be classified into different regions between the ultraviolet region and visible region by the wavelength of light. The ultraviolet region was involved on the wavelength of 200-400 nm, while the visible region was involved on the wavelength of 400-1000 nm, respectively. In this ultraviolet region, the colorless feature was observed while in the visible region, color of light was therefore observed. Figure 35 exhibits the transparent properties of light in ultraviolet and visible light region. The design of cellulose-based composite was opaque in ultraviolet region, whereas it exhibited high transparency in the visible region. Compared to flat glass sheet, the transparency was very high and it was estimated at 100% of transparency. The use of transparent polymer was estimated to be 90% of transparency. In case of cellulose based composite, the transparency was estimated to be 80%. Significant enhancement on transparent properties can be observed due to smoothness of surface of cellulose based composite. Compared to neat nano-scale cellulose sheet, it was opaque and the use of only nano-scale cellulose sheet was unsatisfied in terms of transparency. In case of cellulose-based composite, significant enhancement on light transmission was therefore observed. It offered high transmittance in the range of 75–87% in the visible light wavelength (400–800 nm). Such high light transmittance satisfied the transparency criteria for the OLED substrate of 80%, suggesting the cellulose-based composite gain interest as a potential candidate for flexible OLED substrate.

The transparency of cellulose-based composite was successfully developed due to the increments on the smoothness of surface. The use of polyurethane-based resin also helped to encapsulate the cellulose sheet and protect it from moisture and the surrounding atmosphere. From the fundamental point of view, cellulose is highly hydrophilic, while moisture is known to greatly reduce the lifetime of organic light emitting devices. Effective moisture and air barrier properties are vital to the performance of organic light emitting diodes. Moreover, the role of polyurethane was not only to protect cellulose from any environmental concerns but also the use of polyurethane can reduce the roughness on the surface of cellulose sheet. The smooth surface of the cellulose based composite also contributed to the transparency of the film as the light diffraction at the surface will be uniform. On the contrary, the relatively rougher surface of the bacterial cellulose sheet will make light diffract in a more random fashion.

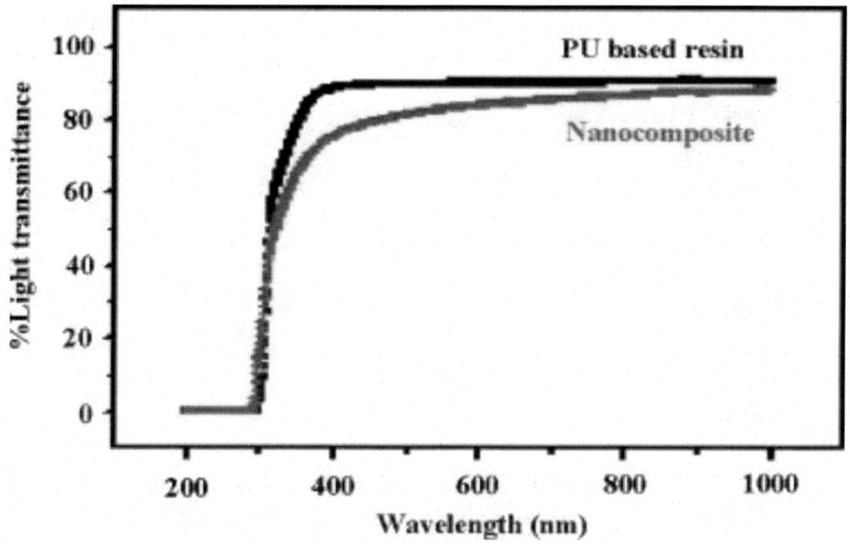

Figure 35. Transparent properties of cellulose based composite [59].

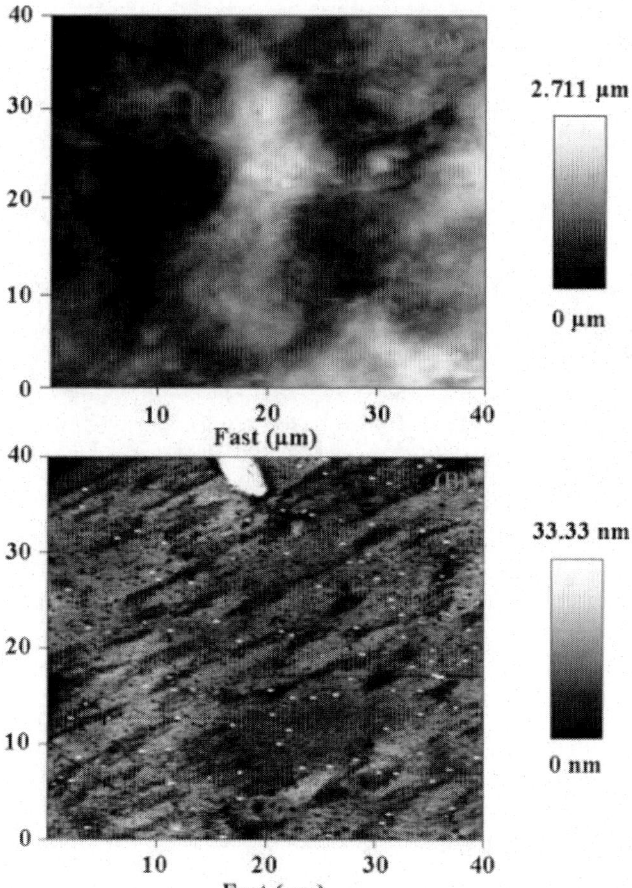

Figure 36. Roughness of cellulose based composite [59].

Figure 36 illustrates AFM images of cellulose sheet and cellulose-based composite taken with lateral contact mode. In this work, five AFM images were taken at different areas on the sample, all of which revealed comparable topologies. The AFM scan size (40 m × 40 m) implied the uniformity of the surface roughness. It can be observed that the smoothness of the cellulose sheet was immensely improved after the impregnation of polyurethane based resin. The sheet roughness decreased from 2.711 m to 33.33 nm as the micron-scale pores of the cellulose sheet was filled with the resin.

Moreover, up to the present time, there have been many investigations on cellulose-based composite for flexible display technology and optical and surface properties. The concern over thermal properties has been extensively realized. Since the application was related to electronic circuits, heat generation was considered one of the issues that could reduce the efficiency and service lifetime of the electronic device.

In general, it was remarkable that only cellulose sheet presented strong interest due to high thermal resistance and low coefficient of thermal expansion. The use of cellulose in flexible display technology therefore gained much interest to be developed as composite. In this case, thermal properties of the neat polyurethane-based resin and cellulose-based composite were analyzed with TGA and DSC. The thermal decomposition behaviors of the polyurethane-based resin and the cellulose-based composite showed that both samples were thermally stable up to 150°C. TGA revealed the polyurethane-based resin had degradation temperature of 265°C while the cellulose-based composite had higher degradation temperature of 345°C. In comparison, pure cellulose had degradation temperature of 370°C. The incorporation of cellulose into the polyurethane-based resin led to an improvement in the degradation temperature, which can be ascribed to the strong interaction between cellulose and the polyurethane-based resin, i.e., hydrogen bonding between the hydroxyl groups in cellulose and oxygen atoms in the urethane unit. On the other hand, DSC revealed that cellulose did not significantly affect the glass transition temperature of the polyurethane-based resin. Both the polyurethane-based resin and the cellulose-based composite expressed glass transition at approximately 16°C. Figure 37 and Figure 38 exhibit the thermal properties of cellulose-based composite investigated by TGA and DSC, respectively.

The dimensional stability of the cellulose-based composite was measured in terms of the coefficient of thermal expansion (CTE). The CTE of cellulose has been reported as 0.1 ppm/K in the axial direction. The low CTE is stable up to 120°C. In this work, cellulose in the sheet form and neat polyurethane-based resin exhibited CTE of 4 and 50 ppm/K, respectively. The cellulose sheet drastically reduced the CTE of composite to an ultimate value of 18 ppm/K, which is below the criteria for the organic light emitting diode substrate of less than 20 ppm/K. This result suggested that cellulose-based composite can be a promising candidate for flexible substrate in display technology.

Another approach to the development of cellulose-based composite for flexible substrate in display technology considered mechanical properties. From the fundamental point of view, in order to determine on the mechanical properties of cellulose-based composite, tensile strength was evaluated. Not only tensile strength but also the evaluation on Young's modulus and elongation at break of cellulose-based composite were therefore investigated in order to predict the feasibility of using cellulose-based composite as a flexible substrate in display technology.

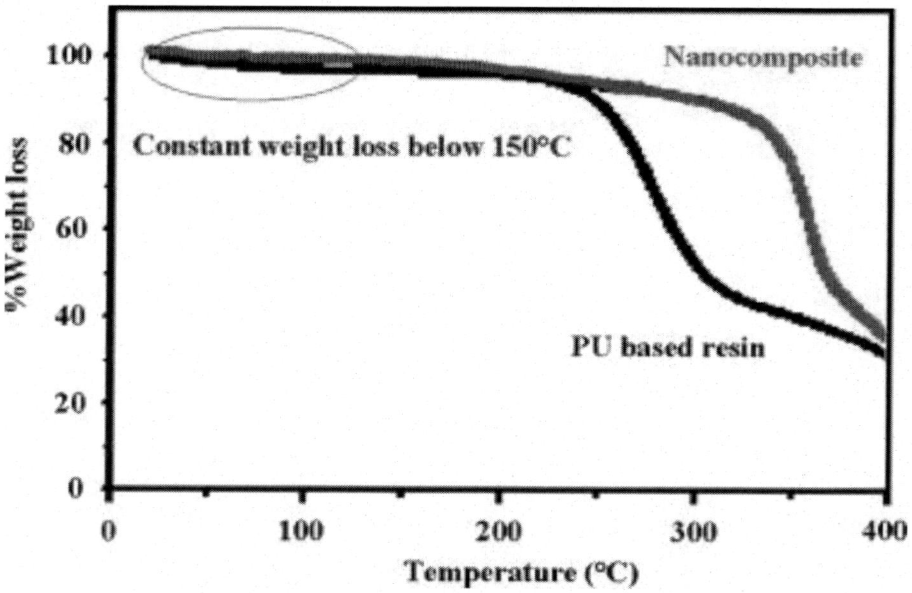

Figure 37. thermal properties of cellulose based composite investigated by TGA [59].

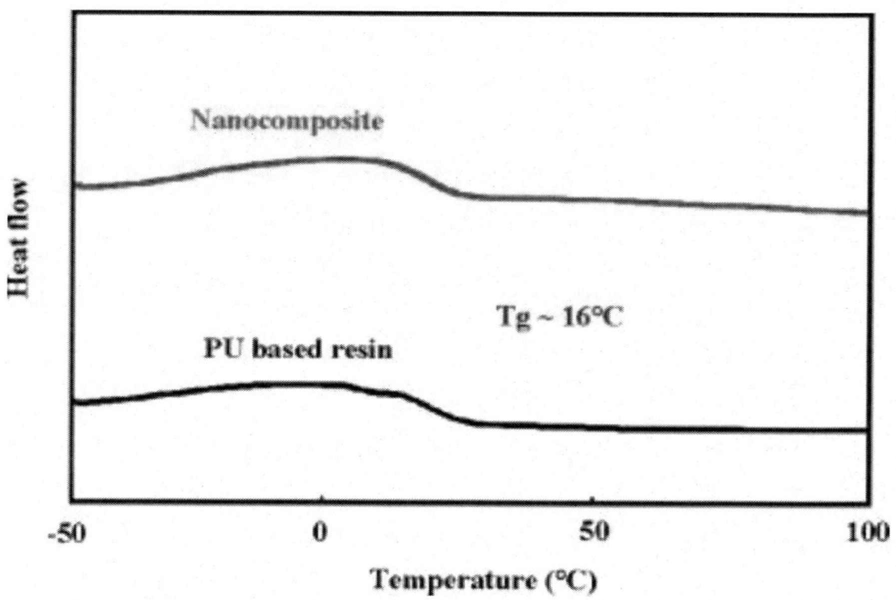

Figure 38. thermal properties of cellulose based composite investigated by DSC [59].

It can be noted that tensile properties of polyurethane-based resin were greatly improved with the incorporation of cellulose. Cellulose-based composite had the highest Young's Modulus and tensile strength, compared to neat polyurethane-based resin. This trend was in accordance with the cellulose contents of the composites. Moreover, it can be found that Young's moduli of cellulose-based composite and the tensile strengths of cellulose-based composite exceeded those of pure cellulose sheet. Cellulose sheet was a three-dimensional network structure of fibers with air interstices in between, when the tensile stress was applied to the cellulose sheet, the stress would be transferred through the fiber network, i.e., the stress could only be transferred 'around' the interstices, not 'through' the interstices. While in the case of the cellulose-based composite, these air interstices were filled with the polyurethane-based resin matrix, enabling the stress to be transferred from fiber-to-fiber through the resin molecules, thus rendering better mechanical properties. Figure 39 and Figure 40 exhibit Young's modulus and the tensile strength of cellulose-based composite, respectively.

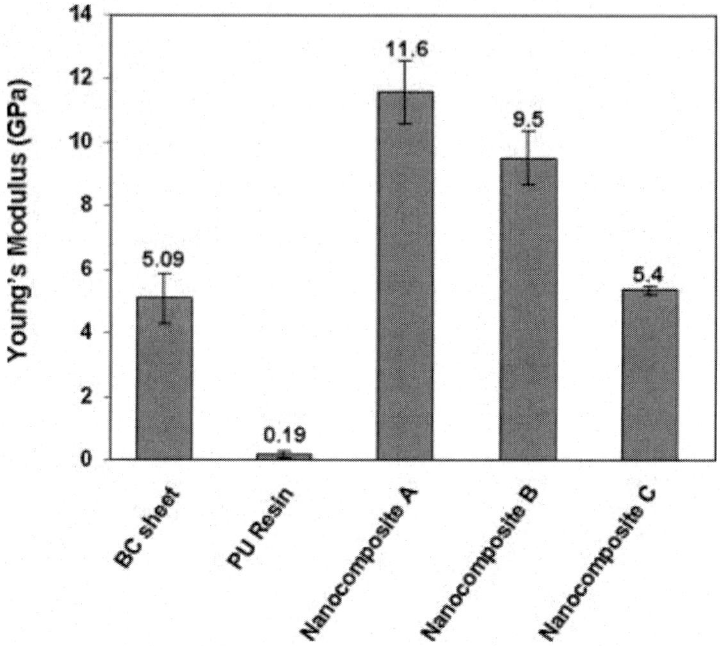

Figure 39. Young's modulus of cellulose based composite [231].

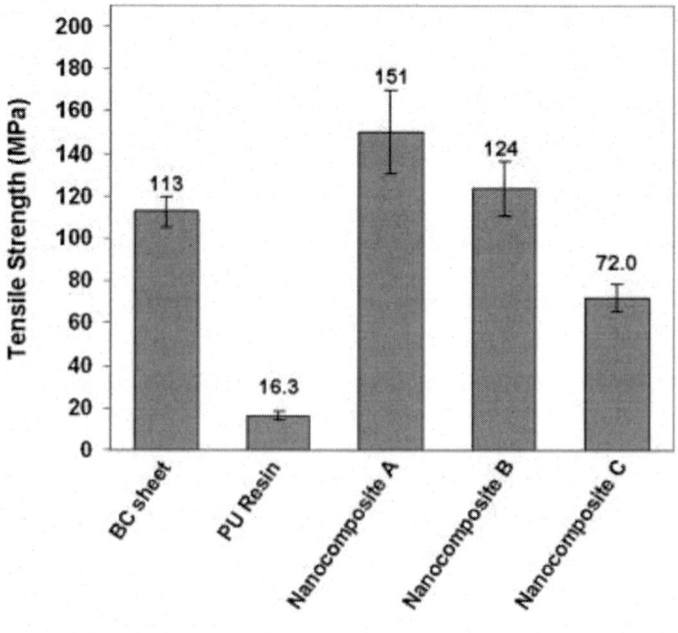

Figure 40. Tensile strength of cellulose-based composite [231].

Moreover, it can be noted that the mechanical properties of cellulose-based composite can be evaluated. Figure 41 shows the graphical representation of the relationship between applied stress and strain derived from the sample deformation. The yield strengths of cellulose-based composite were greatly enhanced with the incorporation of cellulose sheet. The ductile polyurethane-based resin became more brittle by reinforcement of cellulose sheet.

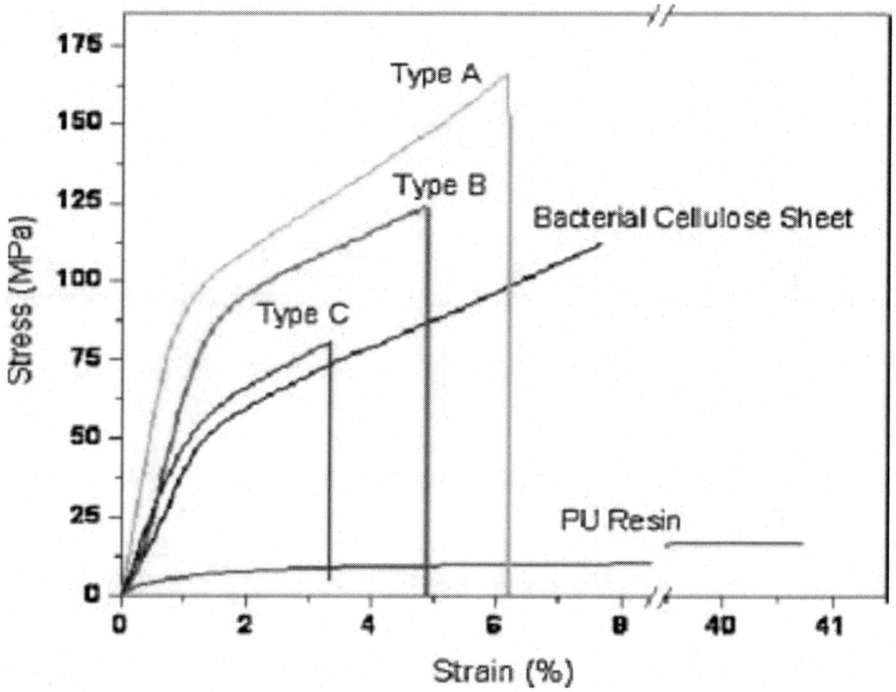

Figure 41. Stress-strain curve of cellulose based composite [231].

3.3.3. Barrier Properties of Cellulose-Based Composite against Moisture

The development of cellulose-based composite as a flexible substrate in display technology has progressed. It can be satisfied on thermal resistance, excellence in optical properties as well as mechanical properties. However, since the prospective application related to the optical device, one of the issues was due to water absorption ability to the electronic device. Consequently, it can lead to decreasing device lifetime, efficiency and quality of device. On the other hand, it can be noted that rapid growth of optoelectronic technology and flexible electronic materials has been evident in the past decade. Among the electronic displays, organic light emitting diodes, with the features of low power consumption, wide viewing angle and excellent contrast, is a versatile platform system that has attracted worldwide attention. Although the organic light emitting diode has traditionally been fabricated on rigid glass sheet substrates, flexible polymeric materials have recently emerged as promising alternatives. Through the achievement of flexible organic light emitting diodes, next generation flexible TV screens, flexible computer screens, and other flexible display systems will be one step closer to our daily lives. Several literatures regarded cellulose-based composite as a potential substrate for flexible displays. Such material features the novel concept of display devices with environmentally friendly aspects. Other benefits include acceptable mechanical properties, transparency, light weight, and ultrahigh dimensional stability in a wide range of temperature. Its flexibility will facilitate the mass production of printed electronics through roll-to-roll (RTR) process. However, one limitation of cellulose-based composite in electronics application is the hydrophilic nature of cellulose. Electronic devices, especially organic light emitting diodes, were highly sensitive to moisture. For an organic light emitting diode lifetime of >10,000 h, the low water vapor transmission rate (WTVR) of 10^{-6} g/m^2/day is required. Water vapor can oxidize the metallic cathode and active organic materials, and thus drastically reduce the lifetime and efficiency of organic light emitting diodes. In order to overcome this issue, various barrier technologies have been investigated [232-234]. The critical challenge lies in the creation of barrier material that offers flexibility and optical transparency. The refractive index of the barrier must also match that of the cellulose-based composite, as mismatching the refractive index will disrupt the transparency of the organic light emitting diode substrate.

One of the most useful techniques for the barrier formation for cellulose-based composite was due to the development of Si-O layer. From the fundamental point of view, it was remarkable to note that this material presented structural properties similar to flat glass sheet and it exhibited very low water absorption ability. In order to gain the benefit from the Si-O layer, optically transparent Si–O barrier film with the thickness in the nano-scale was deposited through plasma enhanced chemical vapor deposition (PECVD). Si–O film prepared through this technique offers the epitaxial crystal growth on the surface with excellent uniformity and adherence to the substrate. PECVD uses ion bombardment on the surface, allowing the possibility of depositing high film quality at low temperature. Consequently the thermal degradation of the polymeric substrate can be restricted to minimal. The preparation of cellulose-based composite substrate is composed of nano-scale cellulose and polyurethane resin.

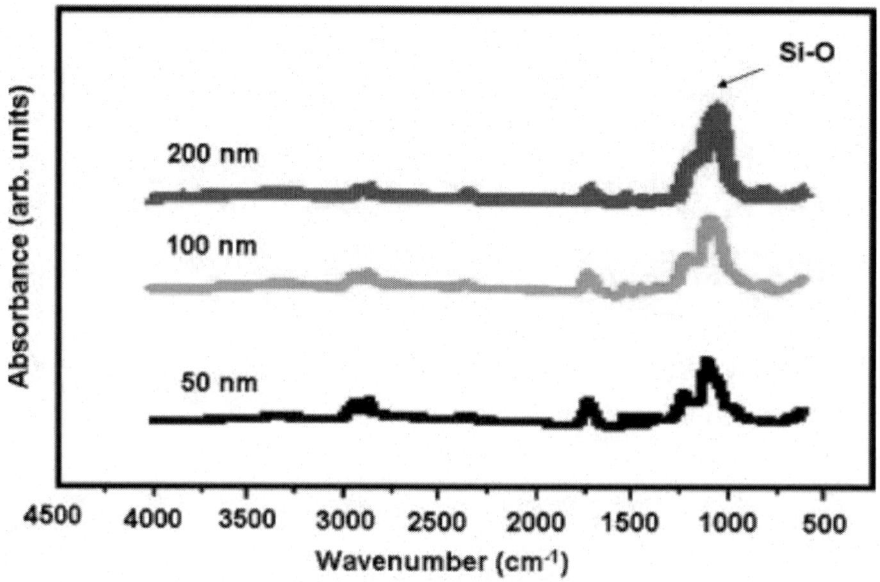

Figure 42. FTIR spectra of Si-O barrier film on cellulose based composite [235].

ATR-FTIR was performed to detect the Si–O layer that was deposited on the cellulose-based composite. The IR absorption characteristic of Si–O has been studied; three major absorption bands centered at 460, 810, and 1070 cm^{-1} have been confirmed by many researchers. These three absorption peaks reflect the rocking of an oxygen atom about an axis through the two silicon, the symmetrical stretching of an oxygen atom along a line bisecting the axis through the two silicon atoms and asymmetrical stretching of an oxygen atom along a line parallel to the axis through the two silicon atoms, respectively. According to Figure 44, absorption peaks centered at 810 and 1070 cm^{-1} can be observed in all samples, confirming the quality of Si–O layer.

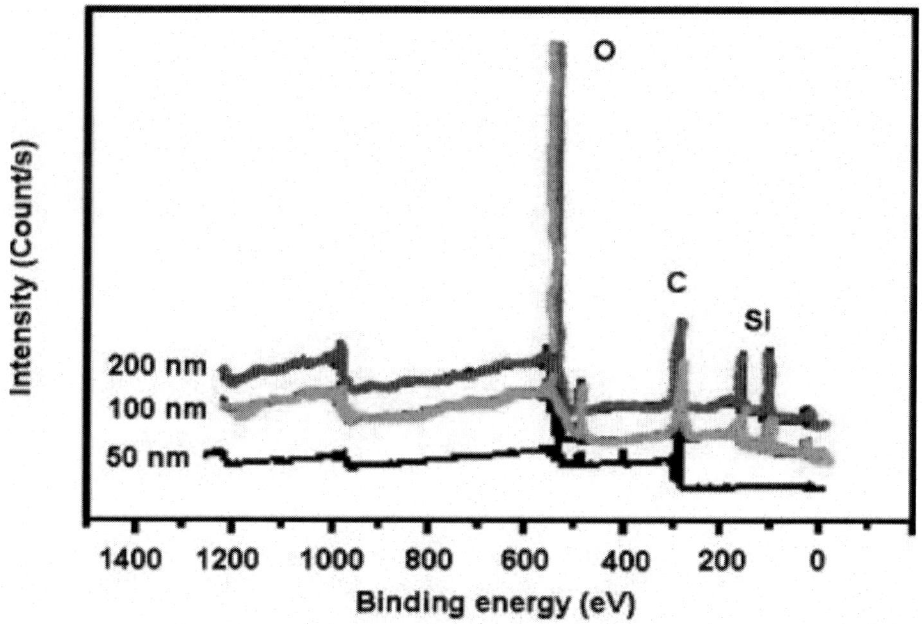

Figure 43. Binding energy of Si-O barrier film on cellulose based composite [235].

XPS was performed to determine the atomic composition of the Si–O barrier film. Low resolution XPS survey spectra revealed that the main elements present on the surface of the deposited cellulose-based composite were carbon, oxygen and silicon (Figure 43). The XPS spectra for ~100 nm-thick and ~200 nm-thick samples show Si peak in the region of 100–120 eV, suggesting the successful deposition of the silicon-based layer. However, the XPS spectrum of the ~50 nm-thick sample does not show the Si peak, suggesting that the deposition time was not sufficient for the effective formation of Si–O thin layer. High resolution XPS was performed in the Si1s region, of which the peaks correspond to the binding energy at 104 eV following XPS spectral line database of National Institute of Standard and Technology (NIST), USA. The peaks revealed that Si content at the surface of ~100 nm-thick and ~200 nm-thick samples was not significantly different, and that the ~50 nm-thick sample did not have Si content (Figure 44).

To investigate the Si–O atomic concentration as a function of depth for all Si–O deposited on cellulose-based composite, the depth profile analysis was carried out by 4 keV Ar^+ etching technique (Figure 45). It can be observed that the surface atomic compositions of the top layer (i.e., at 0 nm depth) of all samples were different from the composition at the adjacent level. Here the 0 nm-depth composition profiles in Figure 47 correspond to the surface concentrations of the Si–O thin film after 1 min of Ar^+ sputtering, which was required for cleaning purposes. In general, prolonged Ar^+ bombardment can lead to significant chemical changes in the film surface. In this case, the Ar^+ sputtering led to a preferential loss of Si and O. The concentration profile of the ~50 nm-thick sample showed no content of Si, which is in agreement with earlier results. The atomic concentration profile consequently belonged to the cellulose-based composite. In the case of ~100 nm-thick and ~200 nm-thick samples, similar and consistent atomic composition profiles of the Si–O bulk layers were obtained. Here Si is in its metallic state without apparent oxidation from the Ar^+ bombardment. Both ~100 nm-thick and ~200 nm thick samples exhibited similar Si concentration of 37–40% and O concentration of 57–59%. The Si and O concentration of the ~100 nm-thick sample dropped drastically as the depth reached 70–90 nm, while in the case of the ~200 nm-thick sample, the drop was observed at the depth of 150–170 nm. Such abrupt changes in the film composition indicated the interface region between the Si–O layer and the cellulose-based composite substrate. Depth profile analysis beyond these interface regions did not detect any trace of Si.

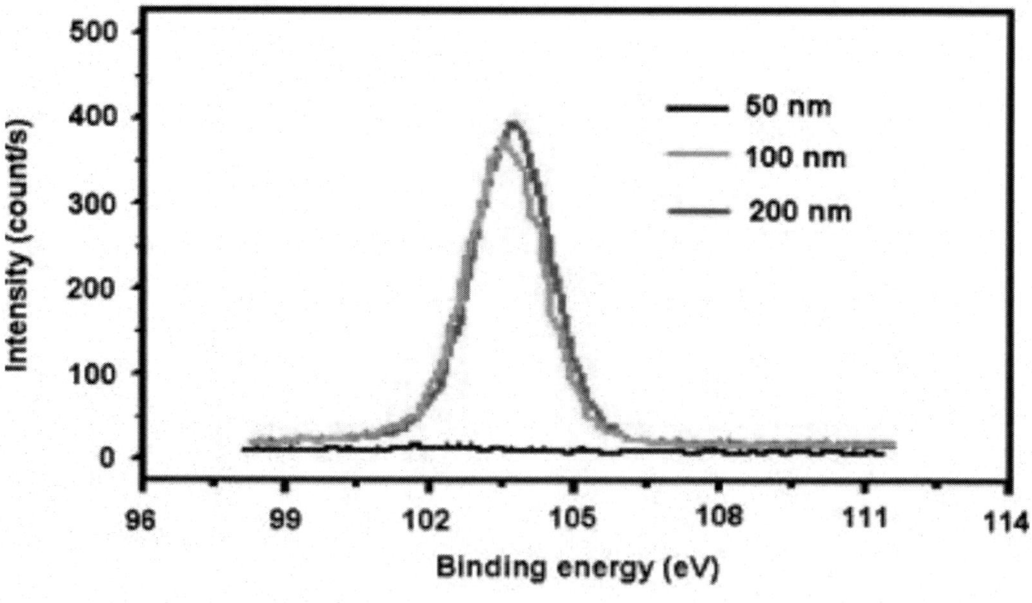

Figure 44. High resolution of Si barrier film on cellulose based composite [235].

Figure 46 illustrates an AFM image taken with lateral contact mode. In this work, ten AFM images were taken at different areas on the cellulose-based composite with Si-O barrier film, all of which revealed comparable topologies. The AFM scan size (25 m × 25 m) implied the uniformity of the surface roughness. It is obvious that the smoothness of the cellulose sheet was immensely improved after the impregnation of polyurethane based resin. The sheet roughness reduced from 1.162 m to 81 nm as the micron-scale pores of the cellulose sheet were filled with the resin.

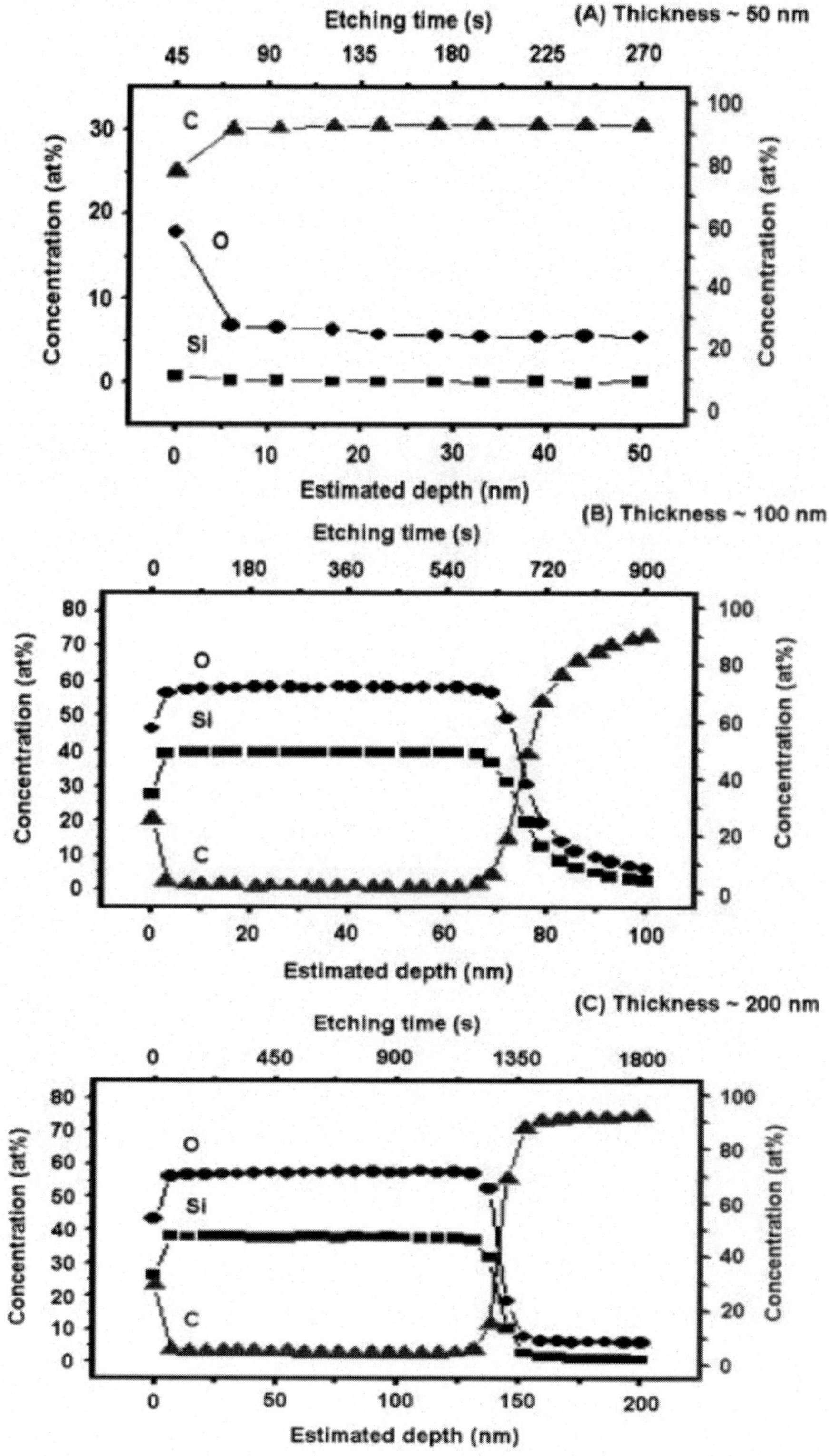

Figure 45. Depth profile analysis of Si-O barrier film on cellulose-based composite [235].

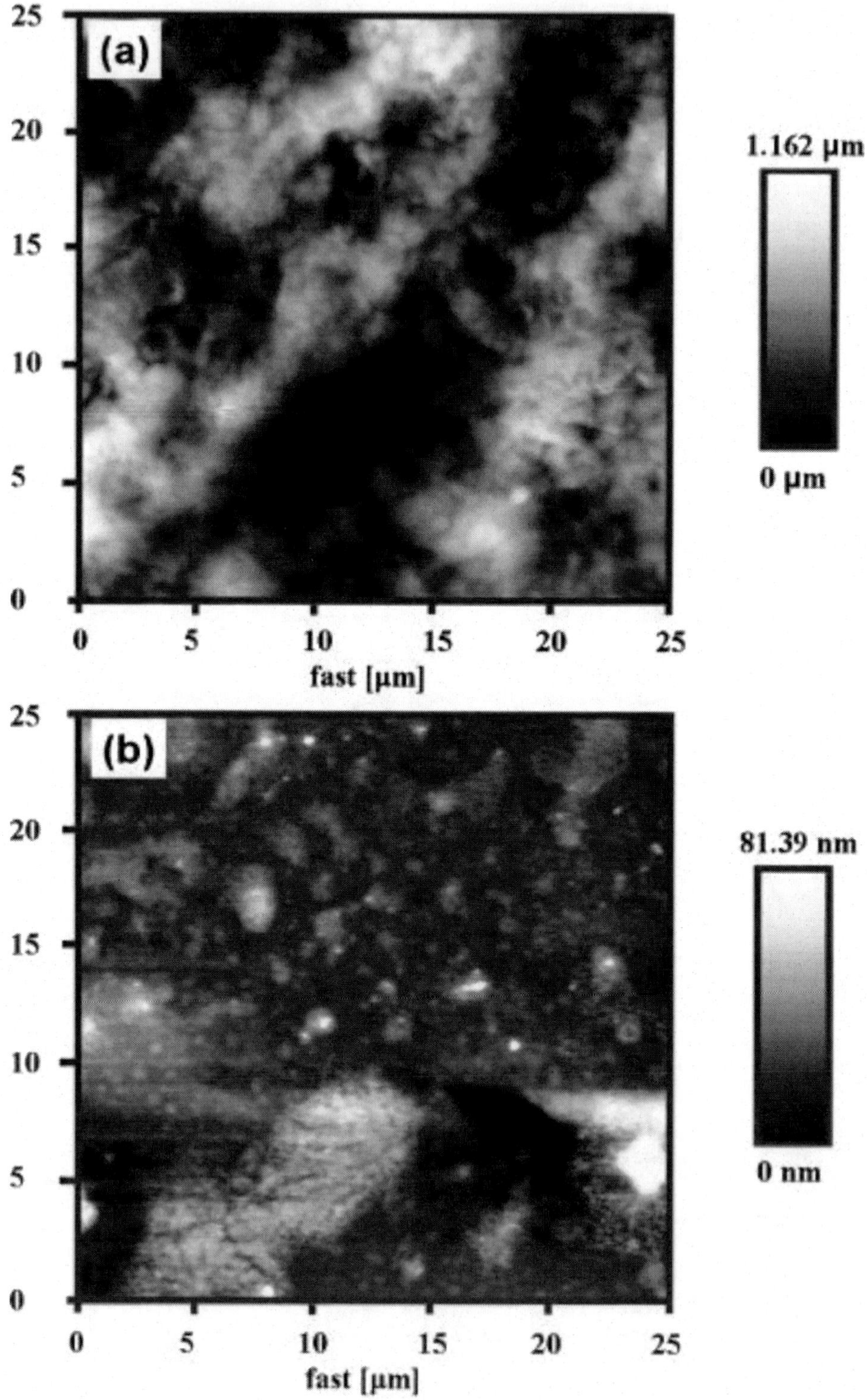

Figure 46. AFM investigation of cellulose-based composite [235].

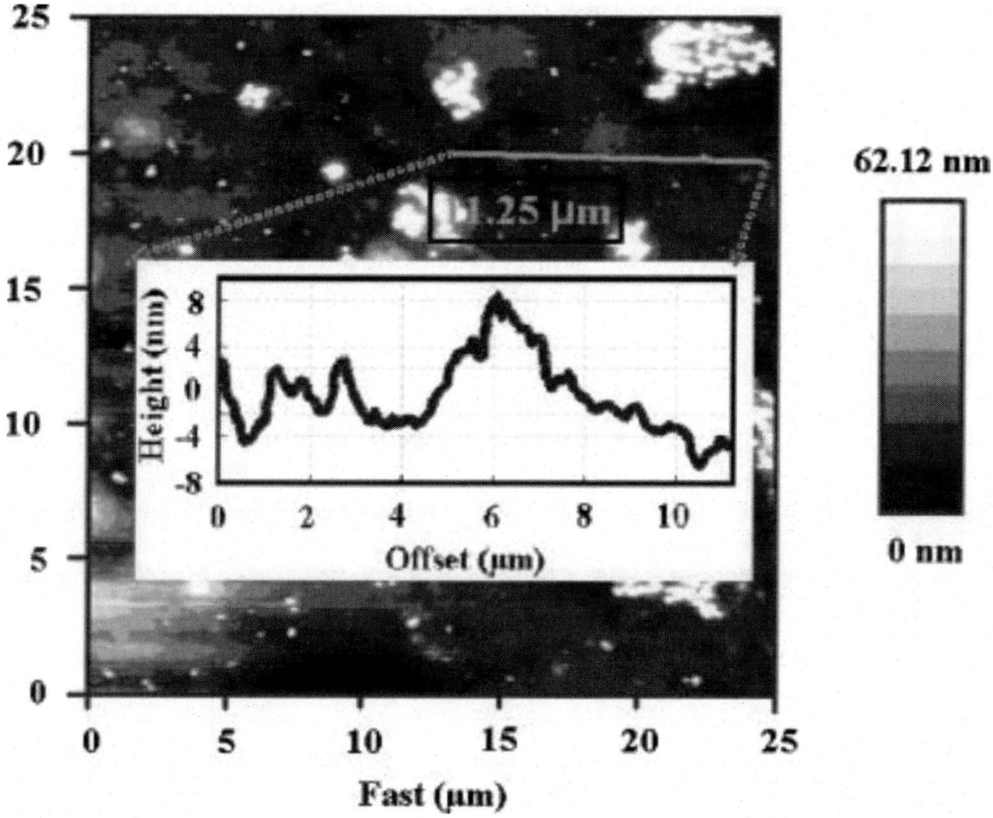

Figure 47. AFM 2D analysis of Si-O barrier film on cellulose-based composite [235].

After the cellulose-based composite was deposited with ~100 nm-thick Si-O layer, the AFM image shows that the substrate surface was covered with Si–O crystallite (Figure 47). The epitaxial growth of the crystallite yielded the uniform topology with the roughness of ~62 nm, which was also a significant improvement over the surface roughness of the pre-deposited cellulose-based composite. Part of the image, indicated by the green line, was sampled from the relatively smooth area, which represented more than 80% of the whole AFM image, for the Fourier transformed pattern analysis – an alternative analysis proposed to facilitate the assessment of 2D surface morphology. This technique involved a statistical analysis to derive statistical parameters, which are correlated with surface roughness. In this case, the length of the green line is 11.25 m and the average surface roughness is ~15 nm, suggesting the exceptionally smooth surface of the greater part (> 80%) of Si–O barrier film. Its roughness was only slightly higher than the specification for organic light emitting diode substrate.

The adhesive force of the Si–O barrier layer was also measured through the FCP technique. The FCP curve for the 100 nm-thick Si–O deposited samples is shown in Figure 50 as an example. In this technique, the AFM tip was brought into contact with the sample while the maximum force needed to pull the tip and the sample apart was measured. As the tip approached the sample within a few nanometers (point A in Figure 48), an attractive force existed between the atoms of the tip surface and atoms of the sample surface. The tip was then pulled toward the sample, and contact occurred at point B. In general, the adsorption of water molecules on the sample surface could also accelerate this so-called "snap-in," due to the formation of water around the tip. As the piezo extended further, the cantilever further deflected, which was represented by the slope portion of the curve. As the piezo retracted, at point C the tip went beyond the zero deflection because of the attractive forces, into the adhesive force region. At point D, the tip snapped free of the adhesive force and was consequently in free air. Therefore, the adhesive force was determined by multiplying the cantilever spring constant (0.58 N/m) by the horizontal distance between point C and D, which corresponded to the maximum cantilever deflection toward the samples before the tip was disengaged. Our Si–O barrier film with estimated thickness of 50, 100 and 200 nm revealed the adhesion force in the range of 116–174 nN.

The organic light emitting diode is extremely sensitive to moisture. In the presence of moisture, the reactive low work function metals used as cathodes can delaminate from the underlying organic layer, while commonly used organic materials can form non-emissive quenching species. The widely quoted requirement for WTVR for an organic light emitting diode lifetime of more than 10,000 h has been reported as 1×10^{-6} $g/m^2/day$. Traditionally, barrier thin films are made of oxide of Al or Si. They are effectively impermeable to oxygen and water. In this work, after the cellulose based composite was deposited with Si–O layer, its moisture barrier performance was greatly enhanced. In comparison to the organic light emitting diode requirement and conventional polymers, the Si–O layer can effectively cover the substrate roughness and fill the pinholes, reducing WVTR down to the order of 10^{-4} $g/m^2/day$. The cellulose-based composite deposited with Si–O layers with the thickness of 100 nm and 200 nm exhibited WVTR performance in the same range, suggesting that water vapor permeation occurred through defects or nano-scale pinholes, rather than the bulk of the Si–O layer.

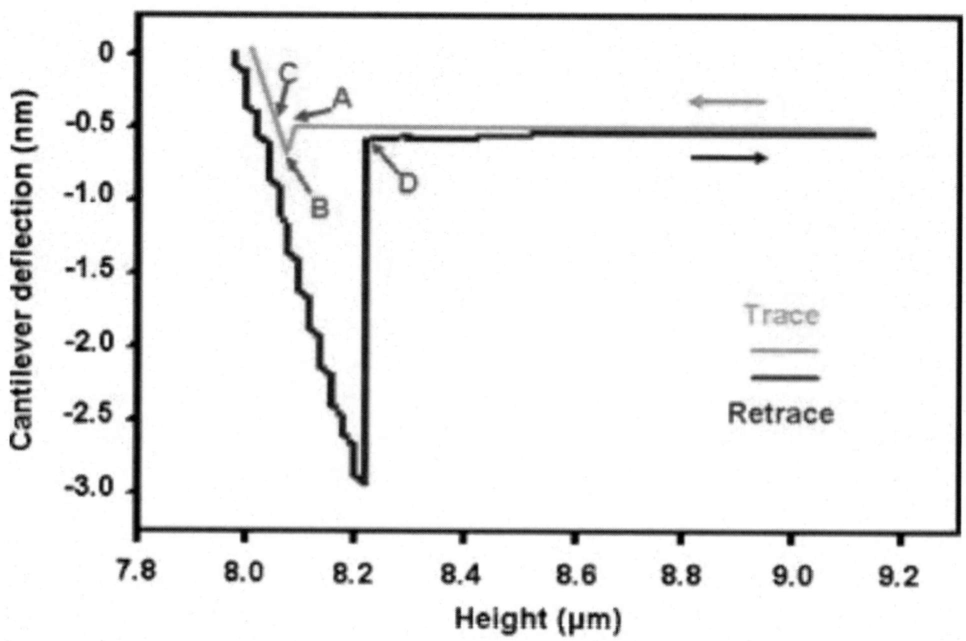

Figure 48. Typical FCP adhesive force measurement on Si-O barrier film on cellulose based composite [235].

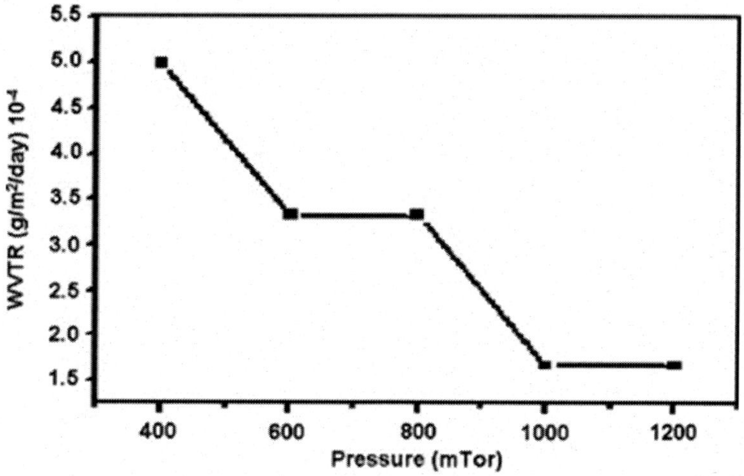

Figure 49. Effect of pressure in PECVD process on WVTR of Si-O barrier film deposited on cellulose based composite [235].

Many governing factors exist in Si–O PECVD, including deposition time, RF power, pressure, gas flow rate and plasma stabilization. These parameters will ultimately affect coating properties such as uniformity, thickness and density. Coating properties can be improved with deposition power, as increasing deposition power will enhance excitation and dissociation of $Si(H_4)$ and O_2 molecules in the plasma. Heat can also improve the mobility of the active species and thus expedites crosslinking of the Si–O network, eliminating the porous defects. As the WVTR performance of the Si–O deposited cellulose-based composite in this work was still two orders of magnitude short of the organic light emitting diode requirement, in the following section, the effects of deposition pressure and RF power on the WVTR performance are investigated. The WVTR of ~100 nm-thick Si–O deposited on cellulose-based composite fabricated at different deposition pressures is demonstrated in Figure 49.

It can be observed that WVTR can be improved by increasing the deposition pressure. Increasing the deposition pressure will decrease the kinetic energy of the active species, as their residence time in the gas phase will be longer, and lead to the formation of Si–O barrier layer with lower residual stress. Consequently, this can present fewer defects, including nano-crack and pinhole. Higher pressure can also induce larger Si–O particles and more stable bonding at the surface of particles.

Figure 50 exhibits the influence of the RF power on WVTR of ~100 nm-thick Si–O deposited on cellulose-based composite fabricated at the deposition pressure of 1000 mTorr. It can be observed that lower RF power is preferred for a higher moisture barrier property. When RF power is increased, a higher amount of active oxygen gas is generated, leading to the formation of the Si–O barrier layer with higher residual stress and therefore more defects. Higher RF power will also lead to a rougher surface. This preliminary work exhibited that Si–O film deposited by PECVD was successfully carried out, and that adjusting PECVD parameters can improve the water vapor barrier property of the Si–O layer. Nevertheless, as the organic light emitting diode is extremely sensitive to moisture, the WVTR is still 100 times higher than the specification for organic light emitting diode substrate. More effective barrier performance of Si–O layer still needs to be developed for the application of organic light emitting diode substrate in flexible display technology.

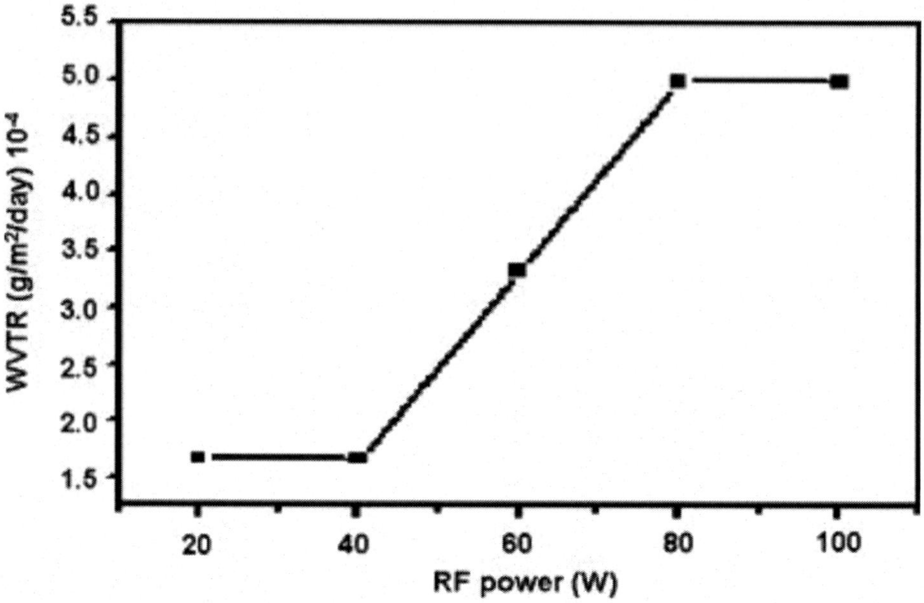

Figure 50. Effect of pressure in PECVD process on WVTR of Si-O barrier film deposited on cellulose-based composite [235].

3.3.4. Development of Smoothness of Cellulose-Based Composite as Organic Light Emitting Diode Substrate in Flexible Display Technology

Next to the development of cellulose-based composite on water absorption ability, the use of cellulose-based composite, as a flexible substrate in display technology, should consider the surface smoothness of cellulose-based composite. This feature of substrate was very important if it was employed as substrate and then the electronic circuit was subsequently deposited on top of cellulose-based composite substrate. In order to solve this issue, the development of surface smoothness of cellulose can be expanded by the use of ferro-fluid solution. Up to the present time, this study on ferro-fluid properties and application has been evident in the past decade [236-240]. From the fundamental point of view, ferro-fluid was considered as a colloidal mixture consisting of magnetic nanoparticles, dispersed in a carrier liquid such as water or organic solvent. It becomes polarized in the presence of a magnetic field. In general, the diameter of magnetic particles in this fluid is about 30 nm and a surfactant is used as stabilizer. Ferro-fluid solution can flow. Its rheological properties can be controlled by applying the magnetic field. Therefore, achievement of this concept can easily lead to the development of biomedical devices, electronic sensor devices as well as controllable devices for agricultural product application.

Among the device applications, the electronic device namely "organic light emitting diode" is a versatile platform system that has attracted worldwide attention. Organic light emitting diode has been traditionally fabricated on glass sheet. The use of polymer substrates has been expected as potential candidate in replacing the rigid and brittle glass substrate in order to achieve the flexible feature. In order to combine the flexible feature and "Green concept", which was defined as the design of the product and process that minimizes the use and the generation of hazardous materials, the use of cellulose-based composite as flexible display is widely employed. By the way, we previously reported the successful fabrication of flexible substrate for organic light emitting diode based on cellulose-based composite. The cellulose prepared from both pulp fiber and bacterial cellulose "Acetobacter Xylinum" provided us an excellent candidate for organic light emitting diode device substrate. Up to the present time, the property of this substrate was continuously developed by Si–O deposition. The use of this nano-scale layer deposited by plasma enhanced chemical vapor deposition (PECVD) can protect the substrate from the surrounding water vapor and consequently enhance the surface smoothness. Therefore, the device lifetime can be extended. However, it has been suggested that, for the effective organic light emitting diode, the surface roughness of the substrate should be kept lower than 5 nm. In order to overcome this issue, several polishing techniques have been investigated. The critical challenge lies in the use of ferro-fluid that offers more smoothness in surface. On the other hand, from the past, the use of ferro-fluid as a nano-scale abrasion polishing technique was well known in both metal and ceramic industries. Figure 51 exhibits the role of ferro-fluid solution on cellulose-based composite substrate.

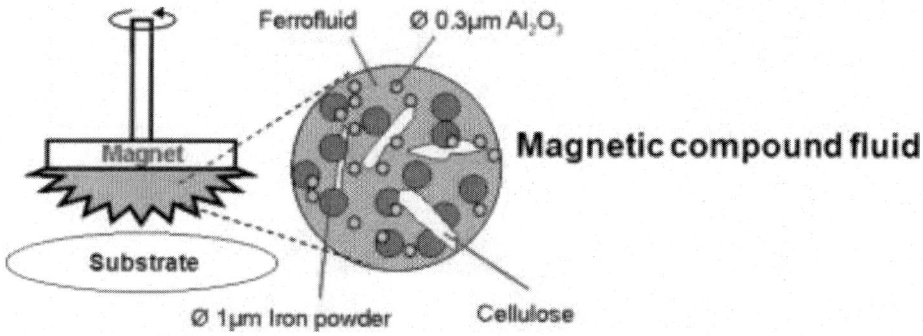

Figure 51. The role of ferro-fluid solution on cellulose-based composite substrate [241].

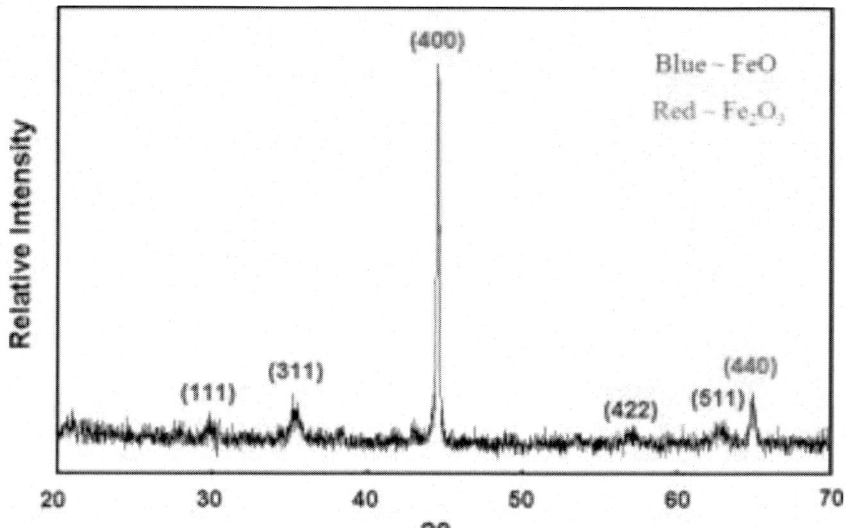

Figure 52. XRD pattern of ferro-fluid solution [241].

The ferro-fluid was successfully synthesized and obtained as a black viscous fluid. After castor oil removal by centrifugation, magnetic nanoparticle was precipitated. It is commonly known that the magnetic nanoparticle is not stable, especially under moisture conditions. The nanoparticle can be oxidized and therefore lose magnetism gradually, which is consequently detrimental to application. To maintain the magnetic nanoparticle stability, the synthesized product was stored in desiccators before usage. Figure 52 exhibits the X-ray powder diffraction of ferro-fluid in solid state. After precipitation, solid nanoparticle was casted on glass slide and subsequently dried overnight. It can be determined that the ferro-fluid nanoparticle was well crystallized.

However, it can be observed that the peak presents two phases. The presence of FeO phase and Fe_2O_3 phase was indicated. However, we believe that the quantity of Fe_2O_3 phase is higher than the FeO phase due to the strong intensity of XRD line (4 0 0), corresponding to the Fe_2O_3 phase. The first phase can be remarkably observed at 44 and 65 corresponding to the face-centered cubic structure of Fe crystal planes (4 0 0) and (4 4 0), respectively. The second phase can also be remarkably observed at 30, 36, 58 and 63 corresponding to the face-centered cubic structure of FeO crystal planes (111), (311), (422) and (511), respectively. In addition, the particle sizes of ferro-fluid particle can be calculated from peak of XRD spectra based on the Scherrer's equation: $D = k\lambda / \beta \cos \theta$ where D is the average particle size, k is a constant taken as 0.9 for calculation, b is the full width at half maximum (FWHM) in radians, k is the wavelength of the X-ray, and h is the diffraction angle. From the Scherrer's equation, the synthesized ferro-fluid has been calculated to be roughly 30 nm at (4 0 0) peak, respectively. This is close to the value deduced from TEM image. TEM microscopy was used to determine the particle size and particle size distribution of magnetic nanoparticle in aqueous solution without magnetic field. Figure 53 represents particle size of magnetic nanoparticle with 50,000 and 100,000X of magnification, respectively. They exhibit the irregular shapes of the magnetic nanoparticle. The mean average particle size and standard deviation were roughly estimated to be 30.6 and 9.8 nm, respectively. In the order of magnitude the theoretical results are in agreement with the result of the XRD measurement, presented in Figure 52.

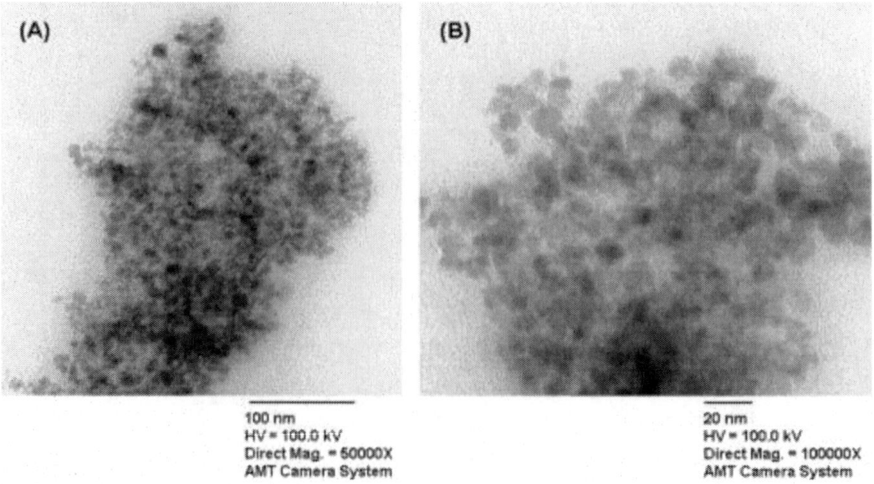

Figure 53. TEM investigation of ferro-fluid solution [241].

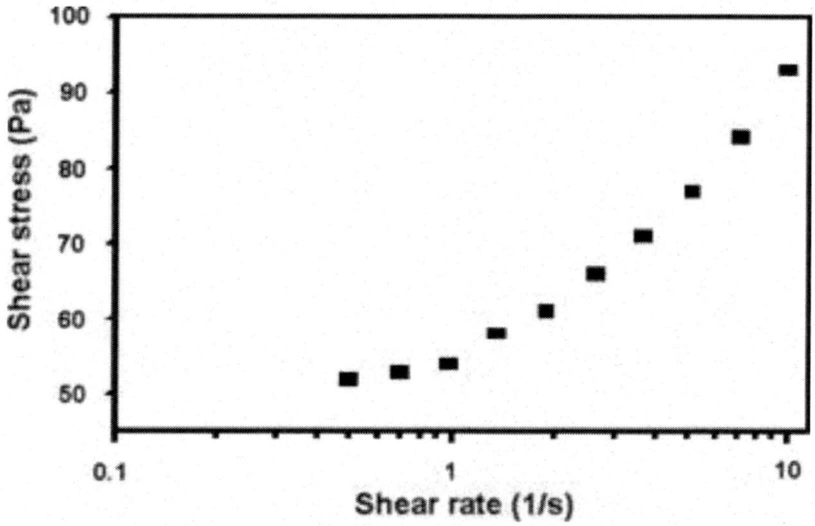

Figure 54. Rheological properties of ferro-fluid solution [241].

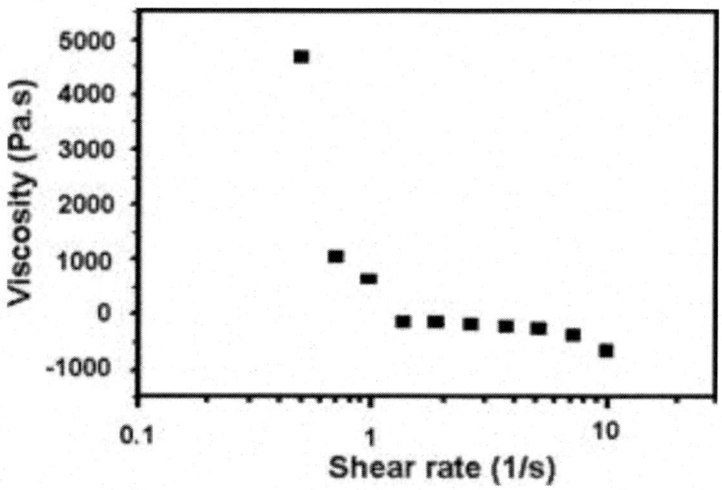

Figure 55. Viscosity of ferro-fluid solution [241].

However, the magnetic nanoparticle exhibited the cluster form having raspberry-like structure in the absence of magnetic field. Moreover, in order to evaluate the flow ability of ferro-fluid solution, the investigation on rheological properties of ferro-fluid were therefore investigated. Figure 54 represents the curve of shear stress versus shear rate. It can be noted that the existence of yield stress is due to the interaction forces of fluid nanoparticles as they initially resist fluid motion.

Moreover, the rheological properties can be investigated in Figure 55 that, the viscosity of ferro-fluid decreases with increasing shear rate, which can be indicated as "shear thinning behavior." It can be explained that when the applied shear rate increases, the nanoparticles begin to arrange their orientation in the shear direction. Furthermore, increasing shear rate destroys the initial bonds existing among the nanoparticles, and as a subsequence, the viscosity of fluid decreases.

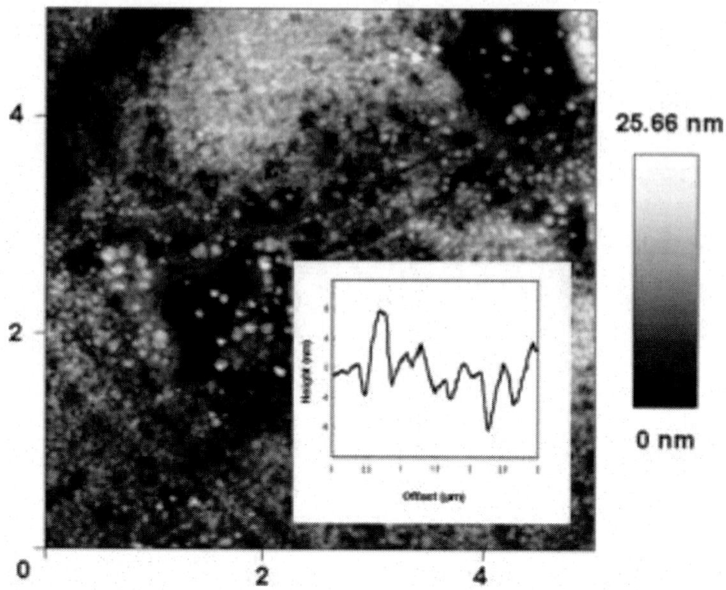

Figure 56. AFM investigation of cellulose based composite after ferro-fluid abrasion [241].

After preparation of ferro-fluid solution, it was used to modify the surface properties of the cellulose-based composite. As a result, the surface of cellulose-based composite becomes smoother. Figure 56 exhibits an AFM image taken with lateral contact mode of a sample polished with magnetic compound fluid A. In this research work, five AFM images were taken at different areas on the sample.

The AFM scan size (5 μm * 5 μm) implied the uniformity of the surface smoothness. According to our previous work, it also reported the smoothness of surface of flexible substrate as 81.39 nm. After polishing by magnetic compound fluid, the sheet roughness reduced to 25.66 nm as scratches on the surface were completely removed; cellulose particles were continually pulled out. In addition, part of the image, indicated by the blue line, was sampled from the relatively smooth area for the Fourier transformed pattern analysis – an alternative analysis proposed to facilitate the assessment of 2D surface morphology. This technique related a statistical analysis to derive statistical parameters, which are correlated with surface roughness. In this experiment, the length of the blue line is 3.12 μm and the average surface roughness is 5 nm. Its roughness was equal to the specification of flexible organic light emitting diode substrate. Moreover, Figure 57 also illustrates an AFM image taken with lateral contact mode. In this work, the AFM scan size (1 μm * 1 μm) was consequently taken. From an industrial point of view, it was suggested that magnetic compound fluid be prepared with a lesser amount of ferro-fluid than proposed. This is probably due to economical reasons. Magnetic compound fluid in condition A used 47 wt.% of ferro-fluid, while condition B used 74 wt.% of ferro-fluid. The data of condition B were provided for comparison. Significant enhancement on surface smoothness can be obtained from both conditions. After polishing by magnetic compound fluid conditions A and B, the roughness was significantly reduced from 26.98 to 3.6 and 4.9 nm, respectively. Therefore, it can be explained that with only 47 wt.% of ferro-fluid in magnetic compound, fluid played an important role on surface smoothness to be less than 5 nm. Our polished surface not only offered an excellent idea from an industrial viewpoint but also reached the specification of organic light emitting diode substrate.

3.3.4. Development of Cellulose-Based Composite for Touch-Screen Properties

Due to growing environmental awareness and concerns over reliable availability of petrochemicals in the future, the polymer industry and academia have been pressed to design and produce polymers and composites based on renewable natural resources. In the past decades, the push towards flexible electronic devices based on "Green composite materials" has been evident. Achievement of this new promising concept can easily lead to flexible displays and optoelectronics, as well as more novel ideas such as smart textiles, photovoltaic cells, and building lighting. Among the flexible electronic displays, the organic light emitting diode is a versatile platform system that has attracted worldwide attention. It has been traditionally fabricated on rigid glass sheet substrates. Although flexible polymer substrates have been expected as potential alternatives in replacing the glass substrate, the use of conventional polymer substrates has been limited by their coefficient of thermal expansion (CTE). It can be noted that the CTE of the substrate should be restricted to 20 ppm/K at most, as the thermal expansion of the substrate can lead to the destruction of functional materials of the organic light emitting diode circuit during the temperature fluctuation in organic light emitting diode assembly and mounting processes. To overcome the CTE limitation of the flexible polymer substrate, our previous works have focused on the exploitation of the composite of nano-sized cellulose and polymeric matrices. Cellulose has CTE as low as 0.1 ppm/K. The incorporation of cellulose into polymeric matrix can be expected to yield a composite film with much decreased CTE of less than 20 ppm/K. In addition, cellulose has the typical thickness and width of 10 and 50 nm. Its nano-entity will allow the fabrication of optically transparent organic light emitting diode substrate as any element with size smaller than one-tenth of visible light wavelength is free of visible light-scattering. Moreover, up to the present time, in order to prolong the service lifetime of electronic device, this cellulose-based composite was further developed for water vapor transmission barrier and surface smoothness, respectively.

Recently, on the other hand, small amount of carbon nanotube (CNT) incorporated on cellulose-based composite has been re-discovered as a smart material that can be utilized for sensors and actuators, so-called electro-active paper (EAPap) [242-246]. It commonly offered numerous advantages including lightweight, dryness, and large deformation as well as minimal power consumption. Due to this achievement of electro-active properties on cellulose modified by carbon-based material, not only elastic modulus and mechanical properties can be improved, but also, it can further the development of successfully transparent and flexible cellulose composite with the additional feature of touch screen characteristic. However, the structure of cellulose was strongly evident as nano-fibril network. If this cellulose is to be further developed in industrial commercialization, it is important to insert a carbon nanotube in the air gap among the cellulose chain. It may covalently be grafted at the OH position in the cellulose structure, not only for electroactive but also mechanical properties. Figure 58 exhibits the chemical modification of cellulose by grafting carbon nanotube on the hydroxyl position of cellulose.

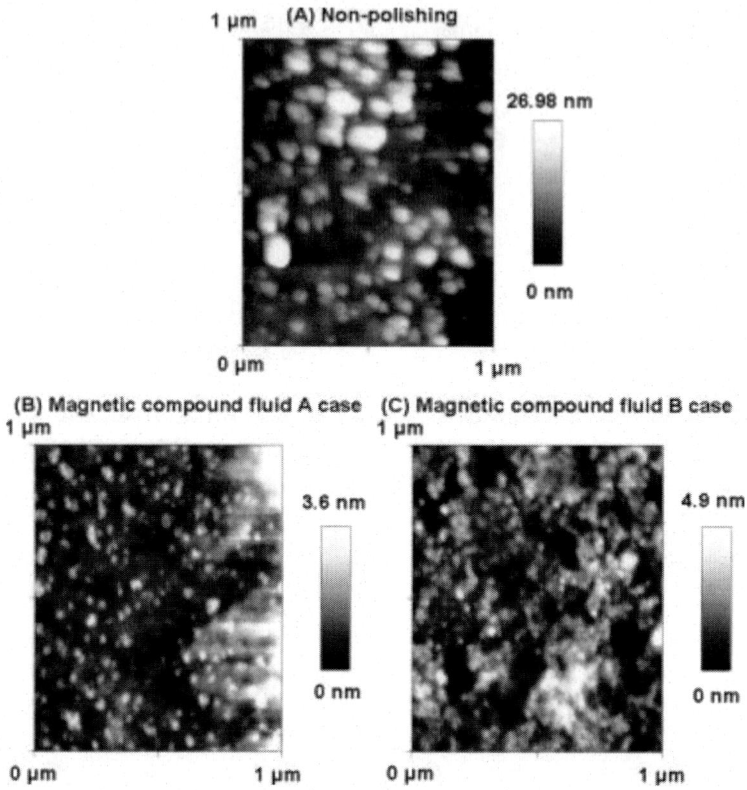

Figure 57. AFM investigation of cellulose based composite after ferro-fluid abrasion [241].

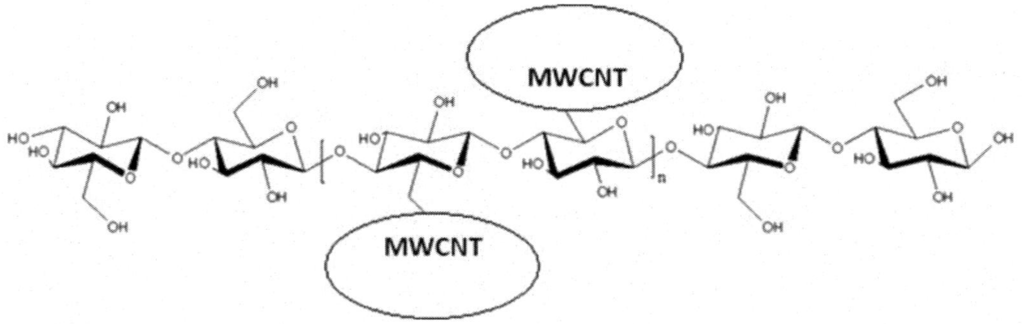

Figure 58. Chemical modification of cellulose by grafting carbon nanotube on hydroxyl position of cellulose.

Carbon nanotube (CNT) grafted on cellulose preparation and characterization was therefore evaluated [184, 247, 248]. It can be seen that cellulose nanofibrils were about 50–100 nm in width and numerous micrometers long. The orientation of exacted bacterial cellulose exhibited three dimensional network structures, with air interstices in between. This is probably due to the fact that along the cellulose molecular structure, there is a hydroxyl group and it resulted in repulsive force acting among cellulose fibrils, disrupting the cohesion force among the cellulose chain. Carbon nanotube (CNT) was well dispersed in deionized water and its particle size was roughly estimated as 30–40 nm. In order to investigate the covalently grafted carbon nanotube on cellulose, sodium hypophosphite was used as reaction catalyst. Figure 59 exhibits the FTIR spectra of the chemically modified cellulose structure. The carbonyl transmission peak at 1700 cm^{-1} and the C–N stretching peak at 1350 cm^{-1} of the chemically modified cellulose can be observed.

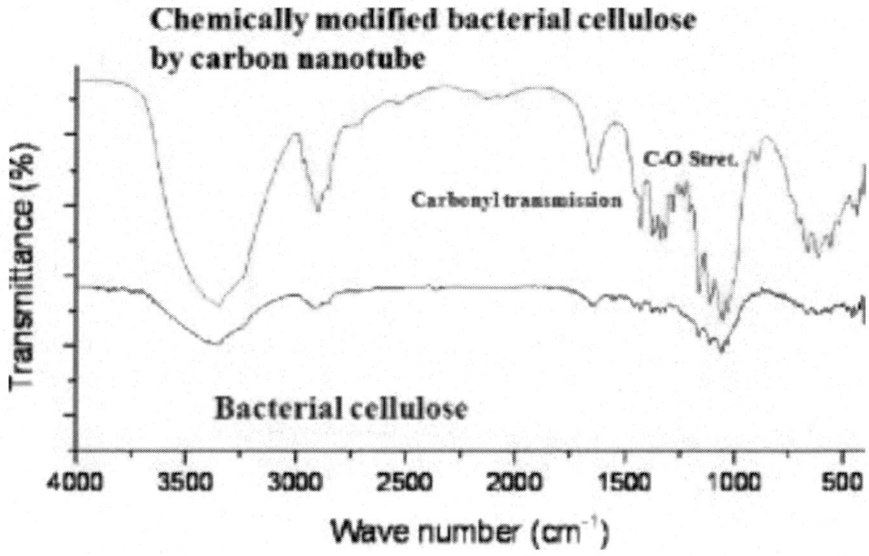

Figure 59. FTIR spectra of chemical modification of cellulose by grafting carbon nanotube [249].

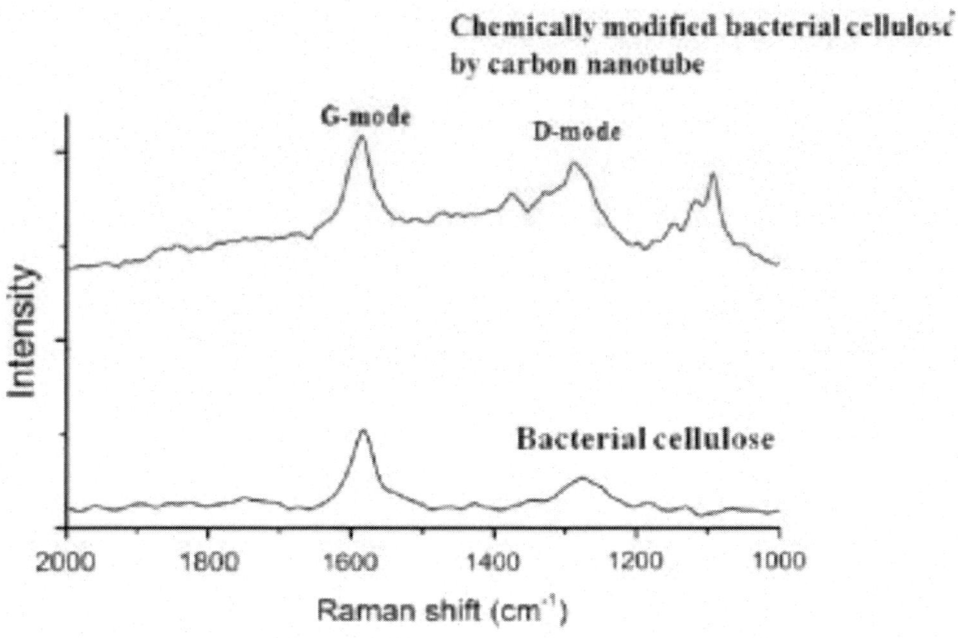

Figure 60. Raman spectra of chemical modification of cellulose by grafting carbon nanotube [249].

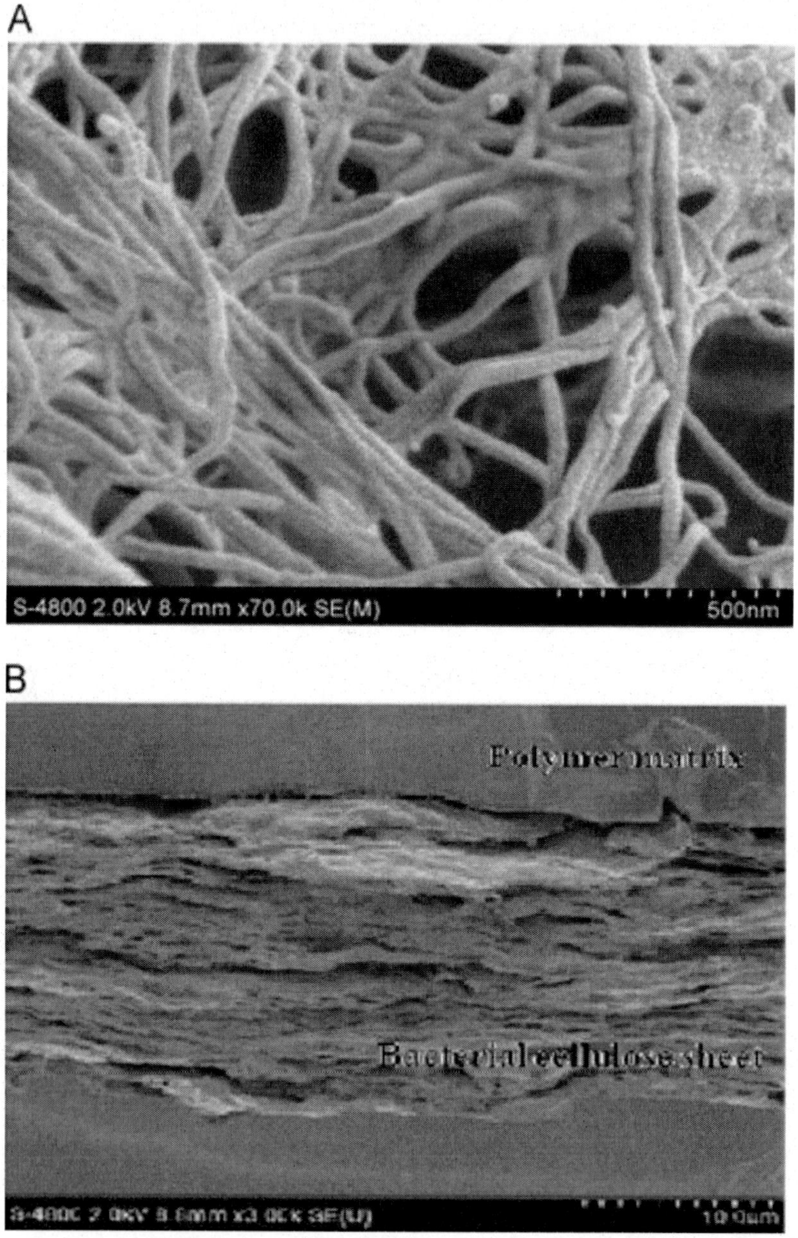

Figure 61. Morphological properties of chemical modification of cellulose by grafting carbon nanotube [249].

The area below these peaks helped explain that the amount of C=O formation from the carbonyl group was superior to neat cellulose, suggesting that chemically modified cellulose was more activated when covalently grafted with carbon nanotube. Moreover, Figure 60 exhibits the Raman spectra of chemically modified cellulose. It can confirm that our esterification reaction by using sodium hypophosphite as reaction catalyst offered the successfully chemical modification on cellulose structure.

The result showed that there were peaks at 1280 cm^{-1} (D-mode) and 1585 cm^{-1} (G-mode), which corresponded to disorder induced mode and tangential mode of carbon nanotube, respectively. However, significant small peak can be observed on carbon nanotube incorporated on cellulose. Small shift peak at 1270 cm^{-1} can be observed and the area below all peaks was superior to pristine carbon nanotube, suggesting that there was significant change on the amount of carbonyl group on cellulose. This is probably due to the fact that novel covalent bond formation between carbon nanotube and cellulose can be formed.

Furthermore, Figure 61 exhibits SEM investigation on morphological properties of covalently grafted carbon on cellulose. It can be found that there is no significant change on morphological properties over whole cellulose network, while chemical structure was changed. It can be noted that carbon nanotube was completely grafted on hydroxyl group of cellulose molecular chain. From the morphological point of view, the use of covalently grafted carbon on bacterial cellulose was expected to be similar to neat cellulose fibril. Carbon nanotube (CNT) grafted on cellulose composite preparation and characterizations: SEM experiment was used to investigate the morphological properties of carbon nanotube (CNT) grafted on cellulose composite. Figure 61 exhibits the cross-section view of sandwich-like structure of bacterial cellulose incorporated as a stack between polymer matrixes. The small particle size of carbon nanotube was attached at carbonyl group of cellulose particle. Here, the characteristic of polymer matrix either covered the exterior surface of cellulose or penetrated within the free space among the bacterial cellulose particle. The role of polymer matrix consequently led to cover cellulose from other environment effects and to enhance the mechanical properties, respectively. This probably implied that the lifetime of cellulose-based composite can be prolonged when this material is used in relevant applications.

Preliminary experiment on dielectric and piezoelectric properties of carbon nanotube (CNT) grafted on cellulose composite: The variation of dielectric properties of covalently grafted carbon on cellulose based composite was evaluated. The frequency dependence (0.1–1 MHz) of dielectric properties at room temperature can be observed. The range of dielectric constant and dielectric loss was approximately due to 10–15 and 0.08–0.12, respectively. Significant enhancement of dielectric properties of covalently grafted carbon nanotube on cellulose was superior to neat cellulose. This is probably due to the fact that covalently grafted carbon on cellulose can lead to strong polarization effect. Appropriate amount of carbon was successfully grafted on carbonyl group of cellulose and behaved as island-like form. The optimization between appropriate amount of covalently grafted carbon and polarization effect has to be well considered in order to prevent the conductive pathway of carbon along the composite structure. On the other hand, in parallel, the dielectric loss was observed with similar frequency range. It can be found that the linear peak characteristic can be observed with increments of frequency, suggesting that polarization effect from covalently grafted carbon on cellulose was stable at low frequency. With increments on frequency, more vibration can lead to more connections among carbon particles and consequently behave as a conductive pathway along composite. From the fundamental point of view on polarization effect provided by carbon particles, significant enhancement on dielectric properties can be consequently observed. It can be found that covalently grafted carbon on cellulose led to be pushed cellulose composite, pioneered as electro-mechanic sensors. In order to satisfy this criterion, a preliminary experiment on piezoelectric properties was conducted. Only small amount of grafted carbon nanotube (0.05 wt%) on cellulose exhibited piezoelectric coefficient as 0.8. Piezoelectric coefficient can be observed superior to neat cellulose based composite. However, this coefficient was still low and it was limited if the touch screen feature was expected to develop. More and more improvement was therefore required to perform in order to use cellulose composite with the additional feature of touch screen with higher efficiency.

Further, the development of cellulose-based composite for touch-screen has been extensive. Recent developments are very different from previous developments. The process of development involved binary blend process between cellulose and electro-active polymer in order to form a binary blend system [250-253]. In recent years, concern over the shortage of petrochemical materials and the ever growing issue of global waste, along with rising oil prices and the pending exhaustion of landfills, has created a demand for research and development of eco-friendly materials. Numerous types of bio-based materials including cellulose, chitin–chitosan, polylactic acid, as well as soy protein have been extensively used to improve engineering properties, in order to replace petrochemical materials. Among the various types of bio-based materials, cellulose, one of the most abundant biopolymers, has received considerable attention because of its potential applications in this environmentally friendly era. Recently, there has been a great deal of research interest in utilizing cellulose for various applications; one example is composite manufacturing because it provides remarkable reinforcing capabilities, excellent mechanical properties, as well as environmental benefits. Significant research on cellulose-based composites has brought about many engineering applications such as automotive parts, adhesives, textiles, infrastructures, medical devices as well as electronic components. In recent years, our research group has developed a cellulose-based composite for organic light emitting diode substrates. This composite is flexible and transparent when produced by the roll-to-roll process. Moreover, in order to reach the minimum requirement as a flexible electronic substrate, this cellulose-based composite has been further developed to improve water absorption ability and surface smoothness. The epitaxial growth of the Si–O film nano-layer and the use of ferro-fluid solution have been employed to modify the cellulose-based composite. The use of thermoplastic polymer substrates has also satisfied fundamental properties in terms of flexibility. The cost and ease of processing have made cellulose-based composites a good candidate for mass production. Numerous thermoplastic polymer substrates have been developed. Polycarbonate and chemically modified cellulose, such as cellulose acetate butyrate are good examples. However, these substrates have not been used due to the coefficient of thermal expansion. In electronic devices, heat generation from electric circuits is an important factor to consider. If heat generation can be reduced, the service life time and efficiency of electronic devices made with these materials could be improved. The use of cellulose-based composites can overcome the limitation of the coefficient of thermal expansion. Cellulose sheets and cellulose-based polymer composites have a coefficient of thermal expansion of 0.04 and 18 ppm/K, respectively. Moreover, to improve the characteristics of flexible substrates, while following environmentally friendly practices, the use of bio-based materials has been extensively employed in manufacturing. Recently, our research group has successfully developed bacterial cellulose-based composite as the flexible substrate for organic light emitting diodes. The advantage of bacterial cellulose is that it is free from wax, lignin, hemicelluloses, and pectin, which are commonly present in plant-based cellulosic materials. From a theoretical point of view, bacterial cellulose is highly crystalline, with a degree of crystallinity of 90%. Moreover, the Young's modulus and specific surface area were 114 GPa and 37 g/m^2, respectively. We wish to extend our research of this substrate to include touch screen applications. The use of this feature is widely employed in many smart phones and portable computers. Numerous research groups have developed cellulose-based composites for this type of application. Small amounts of MWCNT were successfully grafted on the hydroxyl position of the cellulose molecule. This concept was also proven by our research group; carbon nanotubes were successfully grafted at the hydroxyl position of the cellulose structure, which subsequently provided stronger polarization and enhanced dielectric properties. Chemical modification on the hydroxyl position of the cellulose by carbon nanotubes provided many advantages, including homogeneity of the materials and strong polarization, which are fundamental for excellent dielectric properties. However, for the touch screen feature, where high transparency is required, the use of carbon-based materials has presented transparency problems; the opaque characteristic of the carbon hindered light transmission ability. In order to overcome this limitation, the blending of cellulose with electro-active polymers has been considered. The use of poly(vinylidene fluoride) (PVDF), an electroactive polymer, which is highly miscible with oxygen-containing polymers, results in an interaction between the fluorine atoms and carbonyl groups of the counter polymers. Making PVDF becomes easier to perform. In addition, PVDF exhibits outstanding flexibility, chemical stability, and thermal properties. The objective of this research work is to develop a novel cellulose/PVDF blend system for being a novel type of green composite. The structural, morphological, and thermal properties of BC and the PVDF binary blend system were investigated. Preliminary experiments on the dielectric properties were also evaluated. Cellulose was successfully incorporated into poly (vinylidene fluoride), PVDF. The reason for using PVDF was to satisfy the criteria for dielectric property enhancement. This was typically due to electronegativity force from fluorine atom in PVDF, offering polarity. Figure 62 exhibits the XRD pattern of cellulose and PVDF binary blend composite.

XRD characteristic peaks of the cellulose and PVDF binary blend are exhibited in Figure 62. It can be seen that a small amount of cellulose (less than 10%) was successfully incorporated into the PVDF structure. No change in structural properties can be observed, suggesting that a homogenous product was prepared. However, it is evident that the degree of crystallinity decreased with the reduction of PVDF content. It can be explained that PVDF, which is identified as a crystalline polymer, gradually came up as a sharp characteristic peak. This led to higher dielectric properties. On the other hand, a small amount of cellulose did not affect the structural properties of the binary blend, which was almost fully covered by PVDF. The crystal polymorph of the PVDF sheet reflected in the b and a phase is referred to as orthorhombic and monoclinic, respectively. The b-crystalline phase peak exhibited 2h at 20.4, 36.7 and 40 while, diffraction peaks exhibited 2h at 18.6 and 27 were assigned to the a-crystalline phase. The XRD pattern of the PVDF97.5BC2.5 film was similar to neat PVDF but it did not have a small peak at the phase of 27. It may assume that the b/a ratio of PVDF97.5BC2.5 were greater than that of the neat PVDF. However, it was observed that the small peak exhibited 2h at 22.6 corresponded to the [2 0 0] diffraction plane, which was used to identify the high BC content in the composites. The intensity of the b and a phase and the ratio of PVDF decreased when blending higher amounts of BC. This is probably because the fiber network interrupted the crystalline formation of PVDF.

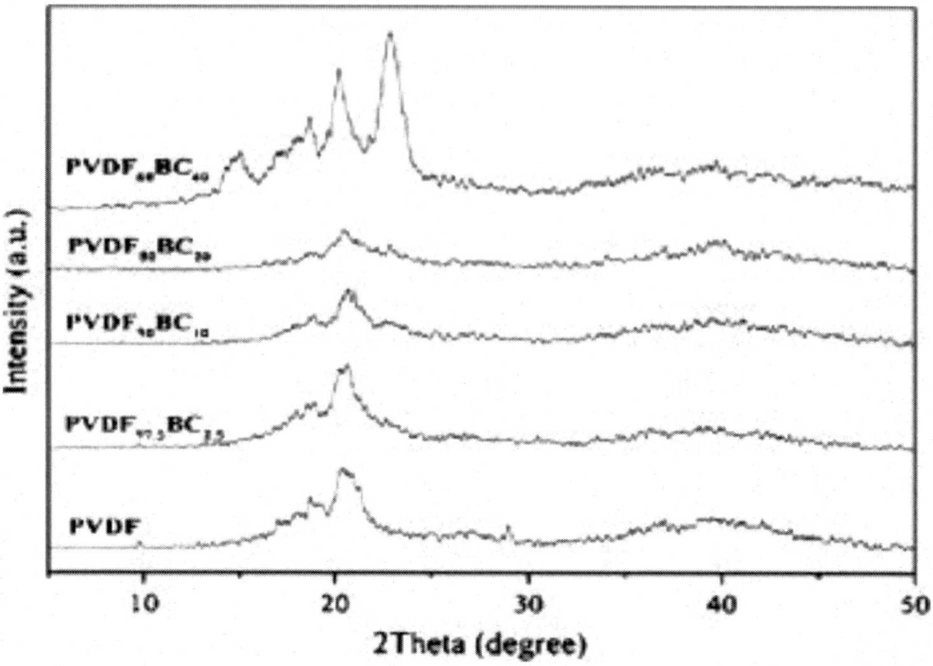

Figure 62. XRD pattern of cellulose and PVDF binary blend composite [254].

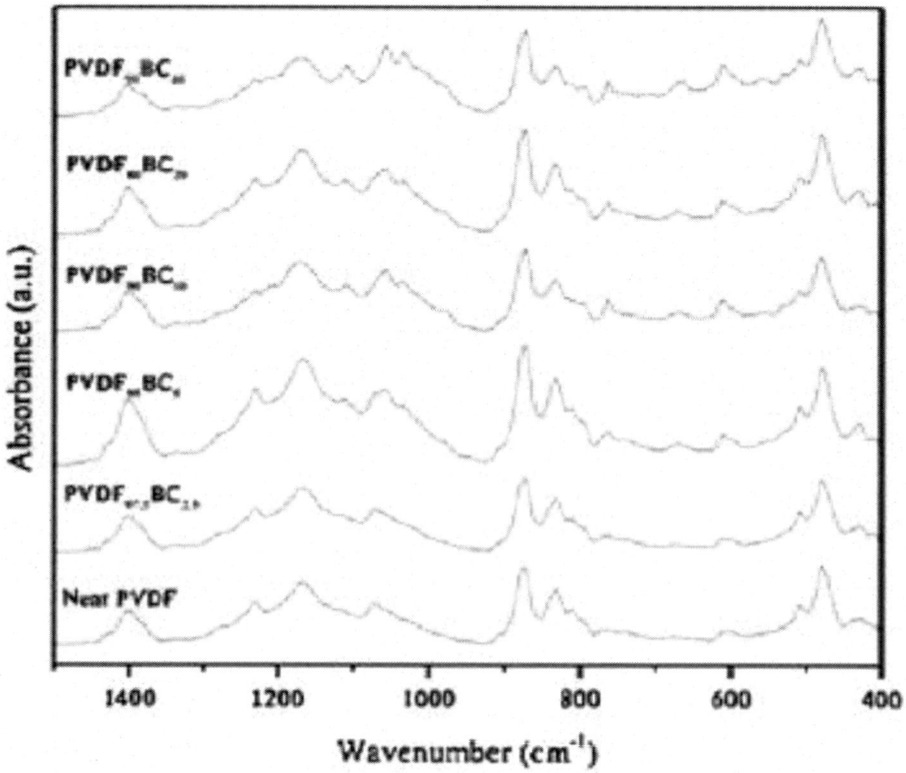

Figure 63. FTIR spectra of cellulose and PVDF binary blend composite [254].

In order to understand the effectiveness of the crystalline phase of the cellulose/be consistent PVDF blend system, FTIR was used to investigate the chemical bonding. The results were tentatively used to determine the amount of each mixture (Figure 63). The peak position of all blend samples was similar to neat PVDF. There was no strong peak of the blend system due to cellulose incorporation. A featureless peak from cellulose appeared. Small peak from cellulose did not affect PVDF structure. Since cellulose did not affect the chemical structure, it may be assumed that only the PVDF phase was sufficiently employed. From a structural point of view, it is commonly known that the chemical formula of PVDF, CH_2–CF_2–, which exists with four molecular conformations (a, b, d, and c), depends on the trans and gauche isomers between the fluorine and oxygen atoms of the polymer backbone. The absorption bands, characteristic of neat PVDF and the composites, demonstrate the two common polymorphs referred to as the a and b phases. The molecule in the a phase shows vibration bands at 530 cm^{-1} (CF_2 bending), 612 cm^{-1} and 763 cm^{-1} (CF_2 bending and skeletal bending) and 795 cm^{-1} (CF_2 rocking), respectively. The formation of the b-phase, all-trans planar zigzag conformations, was confirmed by the presence of specific absorption bands at 510 cm^{-1} (CF_2 bending) and 840 cm^{-1} (CH_2 rocking). The b-phase strongly exhibits dielectric and piezoelectric properties because it shows the strongest dipole-moment, which correlates to the Debye theory. It was needed in order to design the cellulose/ PVDF blend for electro-mechanic response. When higher amounts of cellulose were incorporated, a lower percentage of b phase content was obtained. This was probably due to the fiber network of cellulose disrupting the formation of the b phase of the PVDF matrix. 10 wt% of cellulose was inserted in order to form the b phase of PVDF, which subsequently resulted in better dielectric and piezoelectric properties.

Scanning electron microscopy was used to determine the morphological properties of the cellulose/PVDF blend system. The as-purified cellulose had an average diameter of 30–50 nm and the morphological properties confirmed a nano-fibril network. The size of the cellulose at nano-scale level was well satisfied if it was further developed as a nano-reinforcement in the PVDF matrix. Since our prospective application was to develop cellulose-based composites for touch screen applications, the development of cellulose for dielectric properties and electro-mechanic response enhancement was incorporated in the PVDF matrix.

In Figure 65, both images illustrate how the cellulose network and PVDF penetrated in between the cellulose network. PVDF not only provided significant improvement of dielectric properties, but it also improved the mechanical properties. Since cellulose contains a hydroxyl group in each monomer, the hydroxyl group has a repulsive force, creating space among the cellulose chain. With neat PVDF provided as a comparison, the alignment of cellulose in the PVDF matrix could be seen. The compression technique was used to proceed and consequently offered surface smoothness, and improved mechanical properties, as well as excellent dielectric properties. However, there was a crack phenomenon between the cellulose and PVDF matrix. This may have consequently led to a reduction of the mechanical properties. Prior to application in engineering with any displacement required, the use of a binder or filler needs to be considered in order to prevent mechanical failure.

Furthermore, thermal properties have been evaluated on cellulose and PVDF binary blend system. DSC was used to investigate the glass transition temperature and melting temperature of the cellulose /PVDF blend systems. Figure 68 exhibits the DSC thermo-graph and relevant specific heat parameters, respectively. Within the temperature range of 50°C to 160°C, there are two peaks, which correspond to the glass transition temperature and melting temperature.

The glass transition temperature of neat PVDF, which was provided for comparison, is 40°C. However, the glass transition temperature of the cellulose/ PVDF blend system is slightly increased, at 35°C. The small amount of cellulose incorporated in the system did not affect the glass transition temperature.

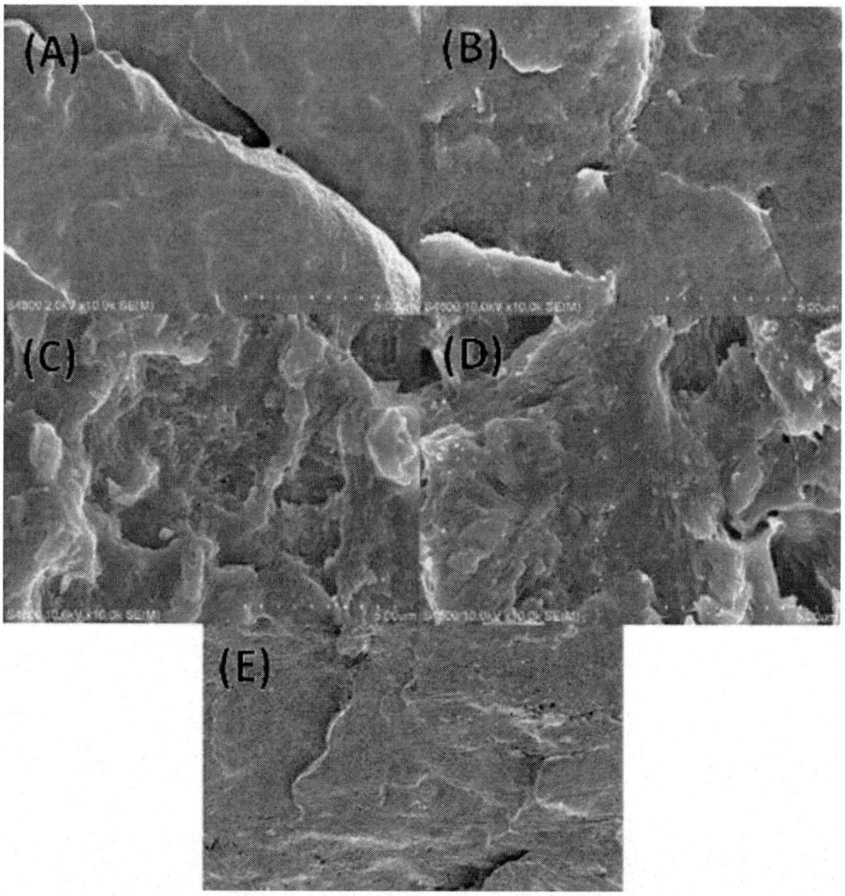

Figure 64. Morphological properties of cellulose/ PVDF binary blend composite [254].

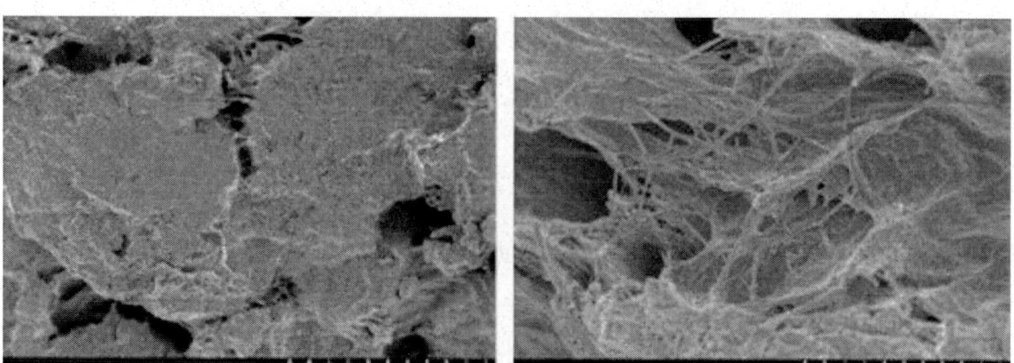

Figure 65. Penetration of PVDF in cellulose network [254].

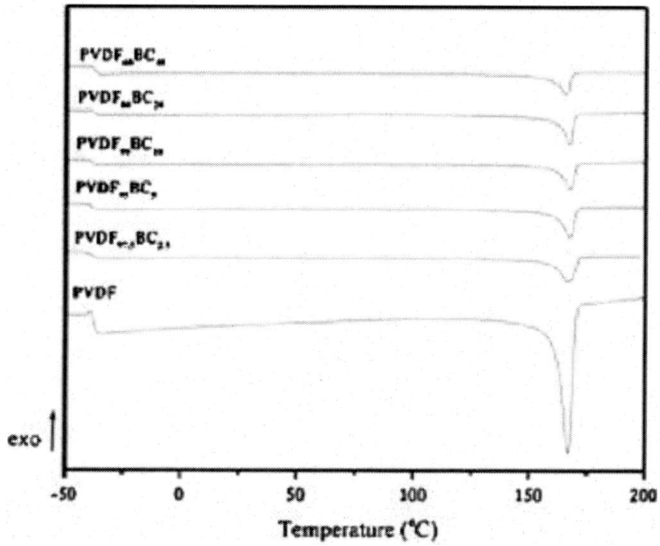

Figure 66. DSC thermogram of cellulose/ PVDF binary blend system [254].

Cellulose did not hinder molecular motion of PVDF from the glassy state to the rubbery state. It was successfully embedded in the PVDF matrix, no evidence of thermal properties was observed. On the other hand, a slight increase of melting temperature was observed. Since cellulose has a high degradation temperature, it may have created a higher melting temperature within the PVDF matrix. Additionally, vinylidine fluoride from PVDF and the anhydroglucose unit from cellulose had an intermolecular interaction and subsequently provided a higher melting temperature superior to that of the neat PVDF. Thermal degradation behavior can be observed by thermo gravimetric analysis (TGA), as shown in Figure 67.

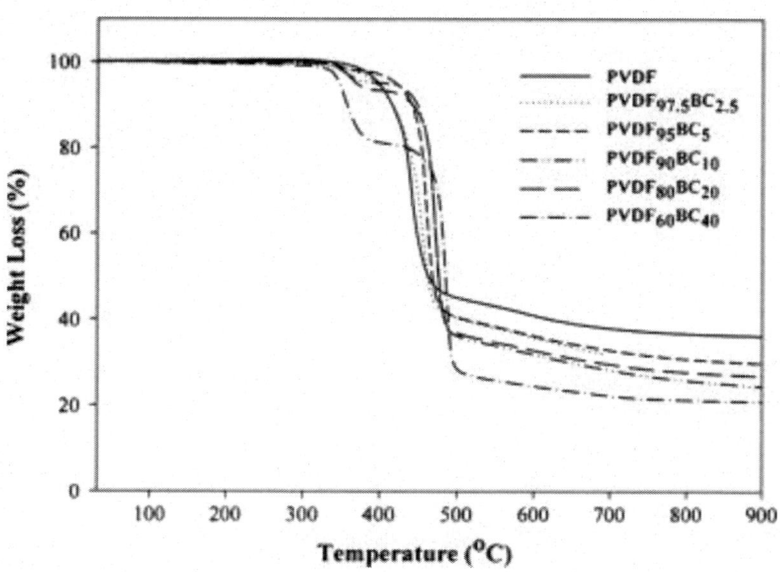

Figure 67. TGA thermogram of cellulose/ PVDF binary blend system [254].

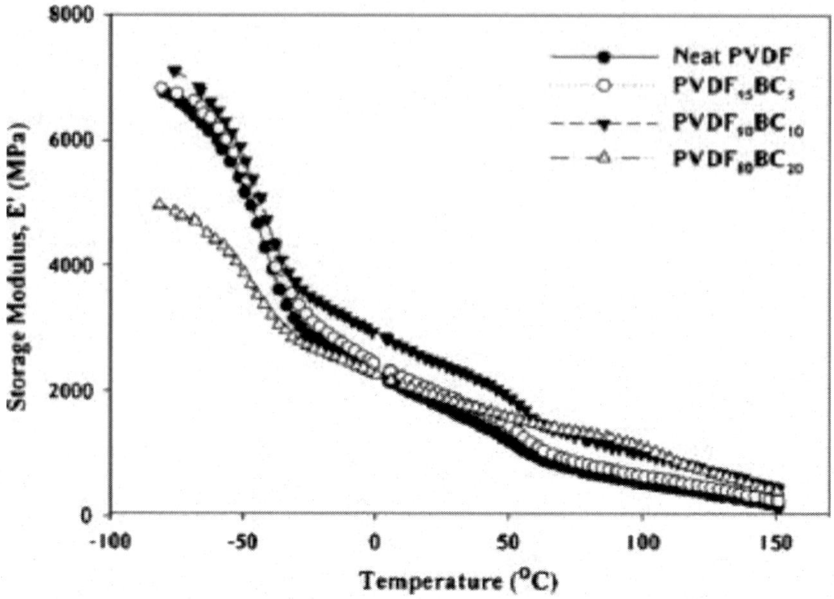

Figure 68. Storage modulus of cellulose/ PVDF binary blend system [254].

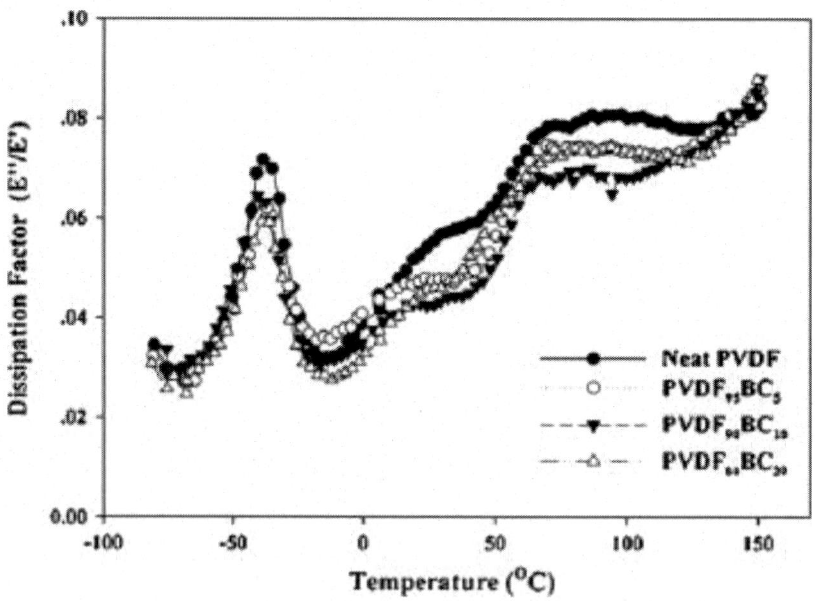

Figure 69. Dissipation factor of cellulose/ PVDF binary blend system [254].

The thermal degradation behavior of the cellulose/PVDF blends can be categorized into two different regions by weight loss. The first region occurs at 320°C and it corresponds to the decomposition temperature of the PVDF matrix, while the second region, which occurs at 460°C, represents decomposition of cellulose. In addition, the thermal behavior of the cellulose /PVDF blend system shows a slight increase over the neat PVDF, depending on the amount of cellulose incorporation. The small increase can be attributed to the H-bond interaction, which is related to the amount of cellulose. Moreover, the cellulose/PVDF blend system is free of DMF, a solvent, and any water absorption. A featureless peak can be observed at 150°C or lower. The thermal behavior of cellulose/PVDF can be determined by dynamic mechanical analysis. In this experiment, dynamic glass transition can be observed over a range of temperatures from 80°C to 150°C.

Figure 68 and Figure 69 exhibit storage tensile modulus and the dissipation factor, respectively. Three relaxation regions can be observed. At 40°C, the relaxation behavior is referred to as the dynamic glass transition temperature. Consequently, at higher temperatures, the second and third relaxation can be ascribed as molecular movement in the amorphous and crystalline phase, respectively. Molecular movement and an interaction between the cellulose and PVDF matrix has occurred. This vibration by applied heat can be explained by changing the status from a glassy-like to rubbery-like state. This is confirmed with the DSC experiment. The glass transition and melting temperature of the neat PVDF, as well as cellulose /PVDF blend system, are 40°C and 150°C, respectively. This is probably due to the degradation region of the PVDF matrix and any segmental movement of its crystalline phase. In addition, if we consider the amount of cellulose in the PVDF matrix, the peak is still similar to neat PVDF, suggesting that a small amount of cellulose incorporation does not affect the thermal properties of the neat PVDF. Cellulose has been successfully embedded in the PVDF matrix. No change in this temperature region can be observed.

The light transmittance of the cellulose and PVDF binary blend system was observed by UV–Vis spectroscopy. The data for the neat PVDF was provided for comparison. The neat PVDF provided a high percentage of light transmittance in the range of 70–90% in the visible light wavelength (400–800 nm). However, compared to the cellulose /PVDF binary blend system, the percent of light transmittance was slightly decreased to 40–60%, depending on cellulose incorporation, suggesting that only the cellulose exhibited opaque properties. Due to the molecular structure of PVDF, the orientation of the fluorine atom exhibited a similar tendency and a high percentage of light transmittance was observed.

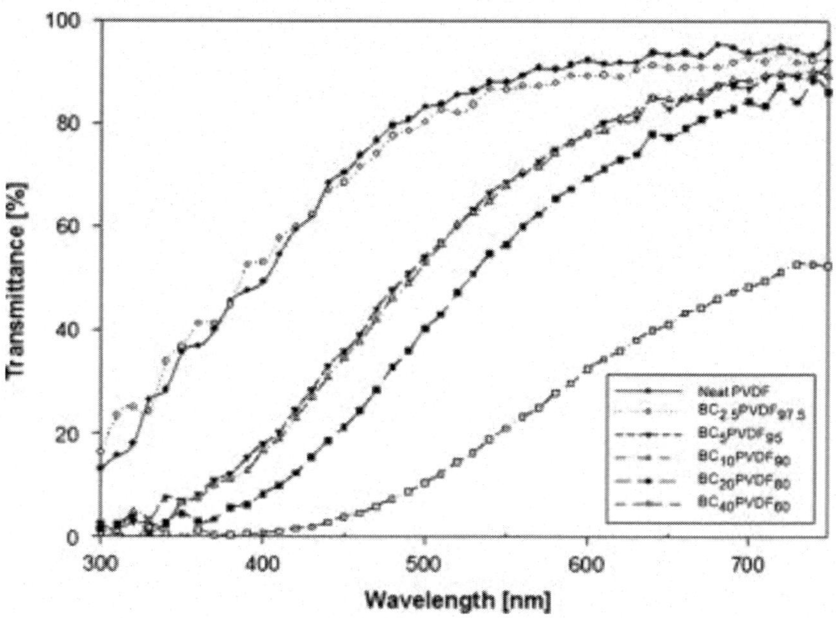

Figure 70. Light transmittance of cellulose/ PVDF binary blend system [254].

Cellulose/PVDF blends are still not acceptable for applications that require a high percentage of light transmittance. More development of transparent properties is required.

Dielectric properties of cellulose /PVDF blend system were therefore evaluated. Figure 71 and Figure 72 exhibit the variation on dielectric constant and dissipation factor, respectively. Preliminary experiments on the dielectric properties and loss tangent of the cellulose/PVDF blend system at ambient temperature were completed. They exhibited frequency dependence on the dielectric constant and loss tangent. The experiment on the neat PVDF was provided for comparison. At low frequency, the dielectric properties were high and they reduced slightly when frequency was increased. This was attributed to relaxation behavior. At microwave frequency or higher, the material did not have enough time to respond and dielectric properties could not be detected. In the cellulose/PVDF blend system, cellulose incorporation significantly improved dielectric properties. This was probably due to the cellulose chain, which contained an H-bond from the hydroxyl group and resulted in greater polarity. If a higher amount of cellulose was incorporated, there was more free space and the repulsive force between hydroxyl groups was reduced. This should be avoided because it decreases the efficiency of polarization, resulting in decreased dielectric properties.

Up to the present time, transparent and flexible cellulose based composite was successfully prepared and used as a substrate for organic light emitting diode in display technology. The cellulose-based composite had appealing features of flexibility, optical transparency, high light transmittance of up to 80% and dimensional stability in terms of CTE of as low as 18 ppm/K. These features satisfied the criteria for the organic light emitting diode substrate. After organic light emitting diode was successfully fabricated on the cellulose-based composite, it exhibited promising performance; it could emit light even when bent.

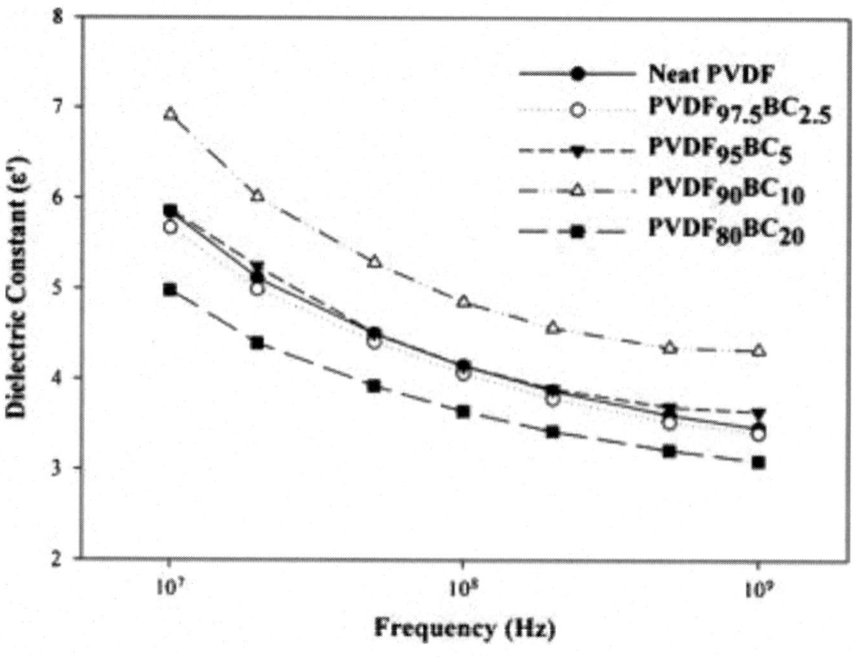

Figure 71. Dielectric constant of cellulose/ PVDF binary blend system [254].

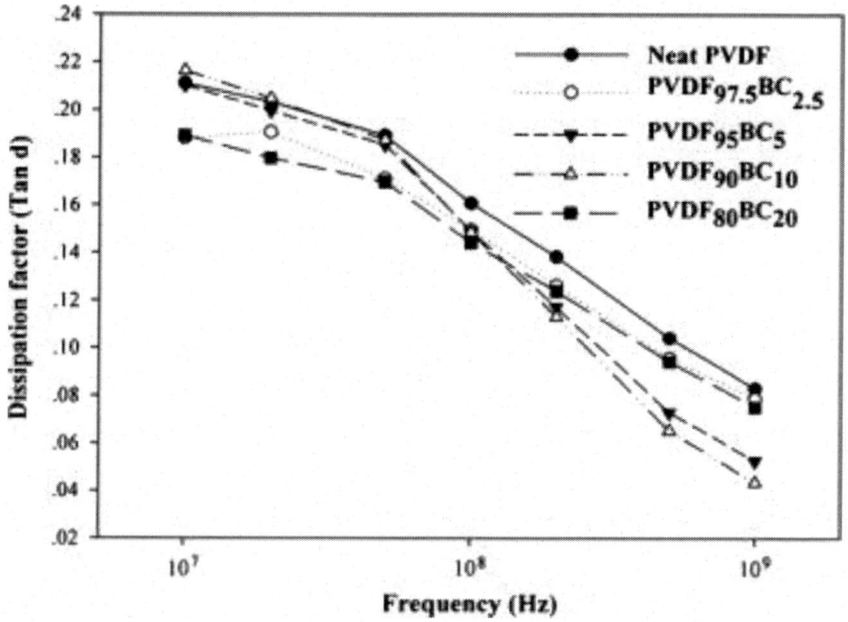

Figure 72. Dissipation factor of cellulose/ PVDF binary blend system [254].

Si-O layer was successfully deposited on cellulose based composite through PECVD. The transparency and flexibility of the composite were retained. The deposited Si-O layer offered a uniform and exceptionally smooth surface. Most of the surface had the roughness of 15 nm. WVTR results revealed that Si-O layer, with the thickness of ~100 nm, had barrier property comparable to that with the thickness of ~200 nm. Si-O deposition can reduce the WVTR down to 10^{-4} g/m^2/day. However, the barrier performance of the Si-O layer needs further development to achieve the specification for organic light emitting diode substrate of 10^{-6} g/m^2/day. On the other hand, ferro-fluid solution was successfully synthesized by the wet chemical synthetic route. It was used as nano-abrasion media for polishing with cellulose based composite substrate. The roughness of the surface was less than 5 nm. Therefore, this experiment can lead us to further develop and scale up flexible substrate of the organic light emitting diode industry.

Furthermore, in order to gain the additional properties of touch-screen, many chemical and physical modifications were therefore employed to investigate its feasibility for being used as touch-screen. A small amount (less than 0.05 wt%) of carbon nanotube was successfully grafted on the hydroxyl group of cellulose and it was further developed as a sandwich-like structure of composite. The insertion of carbon nanotube did not provide the change in cellulose morphology. The excellence of engineering properties including transparent and thermal resistance satisfied the criteria of optoelectronic substrate. In addition, it can be found that small piezoelectric coefficient can be observed; suggesting that incorporation of carbon nanotube can be further developed regarding efficiency if touch screen feature is required. Furthermore, the development on cellulose and PVDF as a binary blend was therefore developed. A cellulose and PVDF matrix was successfully prepared by the binary blend system. From a structural point of view, cellulose exhibited a nano-fibril network and PVDF was used to penetrate the free space in the nano-fibril network. The role of PVDF was to provide polarization. Moreover, the glass transition temperature of the cellulose /PVDF blend was lower than room temperature and preliminary experiments on the dielectric properties found that PVDF incorporated in cellulose as a binary blend could induce the polarization of cellulose. The preliminary experiments on the dielectric properties need further investigation. The binary blend material needs more development of the electro-mechanic response properties for touch screen applications.

Chapter 4

FUNDAMENTAL ON ORGANIC LIGHT EMITTING DIODE

4.1. FUNDAMENTAL ON ORGANIC LIGHT EMITTING DIODE

Organic light emitting diodes are essentially several thin organic film semiconductors sandwiched between two electrodes. A schematic cross-section of an OLED with two organic layers is shown in Figure 73.

The total organic thin film thickness is typically ~ 100 nm. The cathode, top electrode, consists of a low work function metal, typically Al, or Mg: Ag deposited by thermal evaporation. The anode, bottom electrode, is a thin film of the transparent semiconductor indium tin oxide (ITO) deposited onto glass substrate by sputtering. The most commonly used organic materials in OLEDs are: TPD or NPB, as a hole transport layer (HTL), and Alq as an emissive layer (EML) and an electron transport layer (ETL). Their molecular structures are shown in Figure 74. Light is emitted through the anode if the diode is operated under a sufficient forward bias.

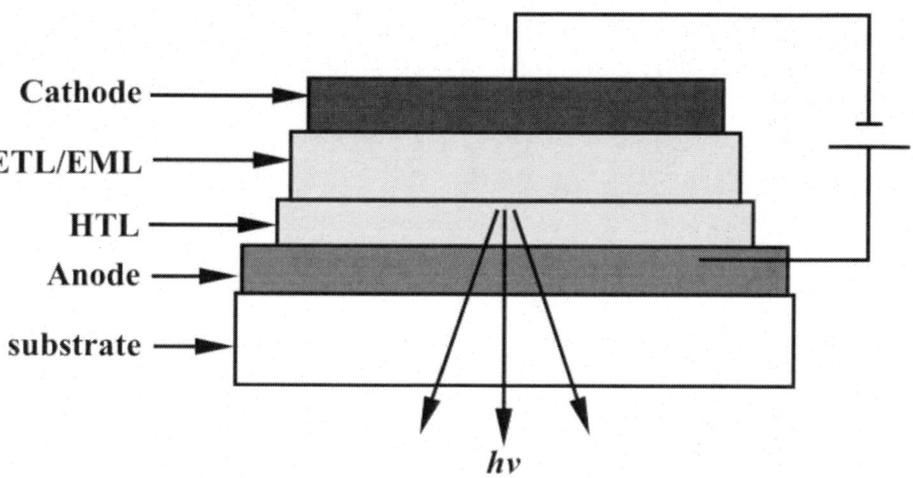

Figure 73. Organic light emitting diode structures.

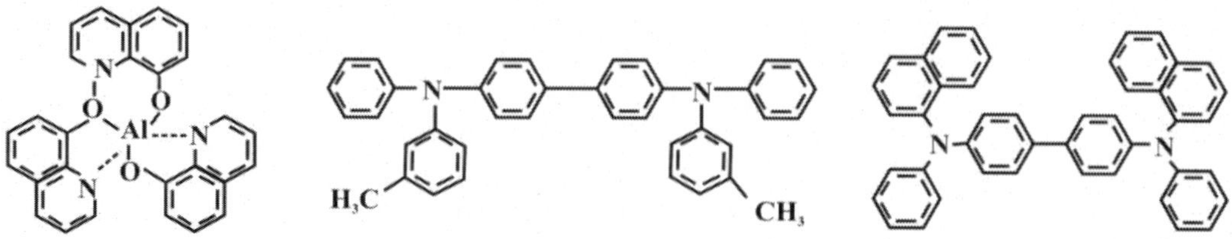

Alq TPD NPB

Figure 74. Structure of some molecular semiconductors that have been used in organic light emitting diode.

Table 4. Properties of organic light emitting diode materials

Material	Glass transition temperature (°C)	Optical gap (LUMO-HOMO) (eV)	Ionization potential (eV)	Electron mobility $(cm^2v^{-1}s^{-1})$	Hole mobility $(cm^2v^{-1}s^{-1})$
Alq	172	2.8	5.7	$10^{-6} – 10^{-5}$	$2*10^{-8}$
NPB	95	3.3	5.2		$8.8*10^{-4}$
TPD	63	3.1	5.4		$> 10^{-4}$

Alq is used as an electron transport and emissive layer, TPD or NPB is used as a hole transport layer.

Device functionality is based on the properties of materials. The understanding of materials is of great importance to further improvement of OLEDs. Considerable research efforts have been aimed at studying the basic properties of those molecules, especially Alq. Some well-documented data are summarized in Table 4. Basically, there are two major concerns. Low glass transition temperature of HTMs such as TPD and NPB is an issue to device reliability. The crystallization of HTM, which can occur at room temperature and can be accelerated by the Joule heat generated during device operation, is one of the important factors resulting in OLED degradation. The other concern is that the electron mobility of Alq is ~ 2 orders of magnitude lower than the hole mobility of HTM. That can lead to unbalanced electron and hole current.

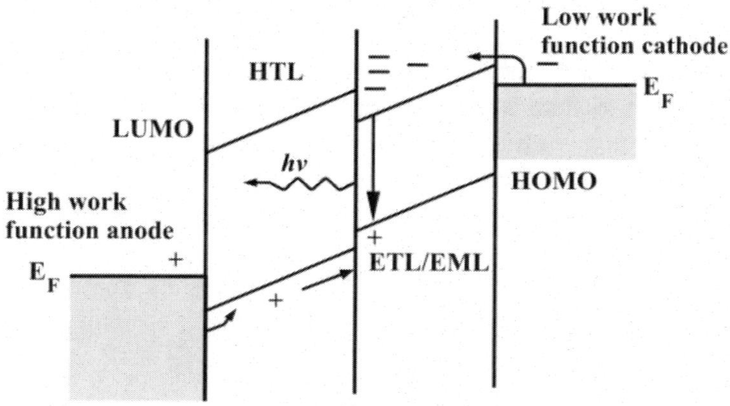

Figure 75. Schematic energy level diagram of organic light emitting diode under forward bias.

Figure 75 illustrates the energy level diagram between interfaces in organic light emitting diodes under a bias, corresponding to the organic light emitting diode configuration as shown in Figure 73. Under a forward bias, holes from the anode are injected into the highest occupied molecular orbital (HOMO) of the HTL, and in the meantime, electrons are injected from the cathode into the lowest unoccupied molecular orbital (LUMO) of the ETL. The recombination of holes and electrons results in photon emission. The radiative recombination occurs at the organic/organic interface, which is generally within the ETL. The organic/electrode and organic/organic interfaces play a very important role in the performance of organic light emitting diodes. Therefore, the understanding of the mechanisms controlling the energy barriers at those interfaces by alignment of the energy levels is critical for good device design.

The external quantum EL efficiency ηφ is defined by the ratio of the number of photons released from the device to the number of charges injected. ηφ can be approximately expressed by ηφ = ηext ηInt, where ηext = 1/2n2 if based on classical ray optics (n is the refractive index of the emissive medium), internal quantum efficiency ηInt is defined by the number of photons produced within a device divided by the number of charges injected. γcap is the factor of the charge balance, defined by γ = Jr/J, where Jr is the current used for charge recombination. The meaning of Jr can be explained by Figure 12 based on the mass balance and charge neutrality of hole and electrons. So two equations can be derived as follows: J=Jh+Je'=Je+Jh,' and Jr=Jh-Jh'=Je-Je.' If all the holes and electrons are consumed for recombination within an emissive layer, γcap will be 1.0. If Jh>>Je or Je>>Jh, γ can be much less than 1.0. Internal quantum EL efficiency can be given by ηInt = γcap · rst · ηPL, where rst is the fraction of excitons, which are formed as singlets, and ηPL is the efficiency of radiative decay of these singlet excitons.

Carrier injection is determined by interfacial electronic properties. The understanding of the formation of energy barriers at cathode and anode interfaces is very challenging. The energy barrier at the interfaces is not exactly the difference between the Fermi level of cathode (anode) and the LUMO (HOMO) of the organic layer since there is an ultra-thin dipolar layer at the interface. However, low work function metals such as Mg, Li and Ca can enhance device performance. Electron injection can play a dominant role in the quantum efficiency of organic light emitting diode if there is a large barrier at the ETL/cathode interface.

Carrier injection into a semiconductor can be treated in terms of either Fowler- Nordheim tunneling or Richardson- Schottky (RS) thermionic emission. Both concepts are appropriate in inorganic semiconductors with extended band states and large mean free paths, which is not the case in organic semiconductors. However, Monte Carlo simulation has shown that this injection mechanism resembles RS thermionic emission even though quantitative differences exist. The rate of injection at a contact limited electrode is proportional to the charge mobility in the organic material. Specifically, the net injected current into the film is the difference between the injected flux and a surface recombination rate. The injection current can be expressed as follows:

$$J_{INJ} = 4\,\Psi^2 N_0\, e\, \mu\, E \exp(-\varphi_B/kT)\exp(f^{1/2})$$

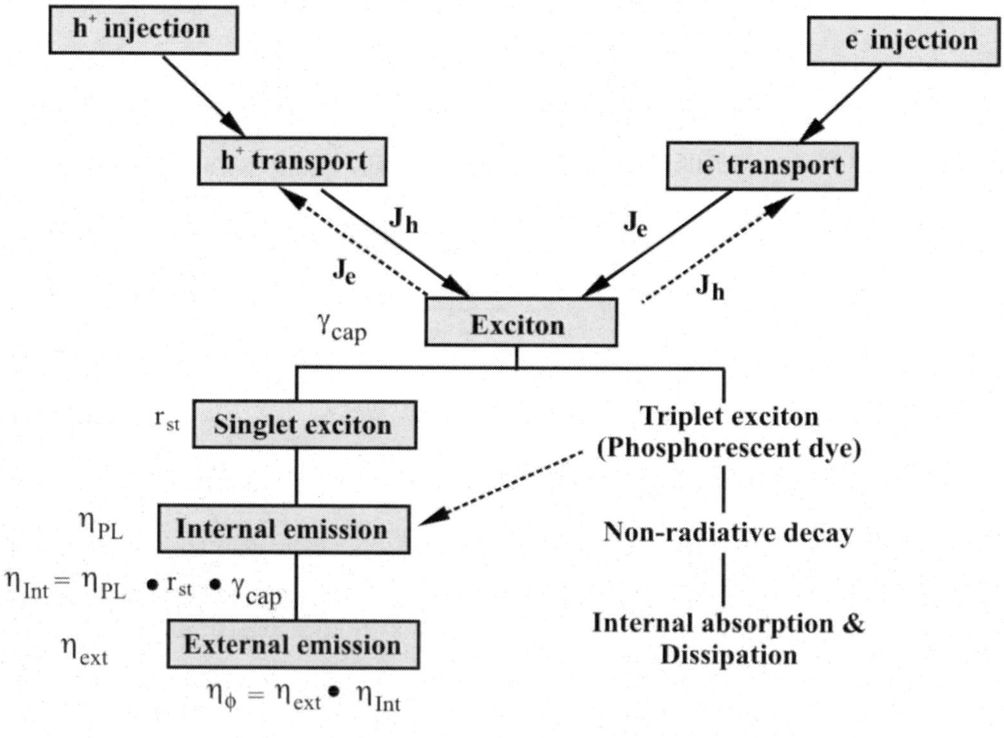

Figure 76. Organic light emitting diode working principles, J_h,' J_e' stand for leakage current in ETL and HTL respectively.

where Ψ is a slowly varying function of electric field, *N*0 is the density of charge hopping sites, and ϕ*B* is the Schottky energy barrier. The exponential in the square root of the electric field, $f = e^3 E / [4\pi\varepsilon\varepsilon 0(KT)^2]$, represents the usual Schottky barrier lowering effect.

Recent intensive studies on cathode interface engineering show that electron injection can be enhanced by introducing less than 2 nm ionic compound films such as LiF, CaF2, Li2O, Cs2O, NaCl, KCl, or by doping the Al cathode in the near interface region with LiF or CsF. The change in image force, determined by the 'f' item in the above formula and caused by inserting the above mentioned high dielectric constant buffer layers, seems to have been overlooked in much of the recent literature. The nature of charge injection is determined by interfacial chemistry, which is more complicated than previously assumed. Also, it is difficult to quantitatively explain the device electrical characteristics on the basis of injection barrier. The understanding of carrier injection also needs a proper knowledge of transport.

Organic light emitting diode materials are conjugated (the bonds between the carbon atoms are alternately single and double). The π electrons are completely delocalized along the conjugated carbons. Hence, their charge transport properties can be affected by the π - π* molecular orbital overlap. However, the overlap of molecular orbitals for intermolecular electron exchange is much smaller than their inorganic counterparts due to weak van der walls bonding between molecules. Physically, carrier transport in amorphous organic materials depends on the electron-phonon and electron-exchange interactions. It is very challenging to disentangle their intrinsic transport properties directly from the electrical characteristics because of a complicated interfacial situation and unknown trap states. In OLEDs, carrier transport strongly depends on the electric field distribution. There is no universal transport mechanism in a wide temperature range. Phonon assisted thermionic emission might be a dominant mechanism at room temperature (RT). At low temperatures, tunneling can be dominant for carrier transport between molecules. Carrier transport in OLEDs has been described by space charge limited current (SCLC) theories at RT if charge injection is not a limiting factor. SCLC obeys the Mott-Gurney equation:

$$JSCLC = (9/8)\ \mu\varepsilon\varepsilon_0\ (V^2/d^3)$$

where μ and ε are mobility and dielectric constant of the material, *ε0* is the permittivity of vacuum, *d* is the distance between the contacts, and *V* is the applied voltage.

The space charges can be either free carriers or trapped charges. The early observation of the temperature-dependent power law relation of current-voltage suggested a trapped charge limited current (TCLC) model, which employed band models with a distribution of trapping levels below the conduction band. It was developed for bandlike transport rather than hopping transport. Furthermore, constant charge carrier mobility is required, which is contrary to the field-dependence of mobilities. The nature of trapping needs to be clarified for proper device modeling. Thermally stimulated currents (TSC) and thermally stimulated luminescence (TSL) have been used to investigate the trap properties of Alq. The results derived from those spectra are not in good agreement. A trap depth from 0.05 to 0.7 eV and a distribution of trap states from 0.13 to 0.25 eV are suggested by TSC and TSL spectra respectively. TSC spectra vary strongly for Alq of different suppliers. Trap states are more like an extrinsic factor. Very recently, non-dispersive electron transport in well-purified Alq film at nitrogen atmosphere also indicates that trap states might be attributed to impurities and oxygen. Trapping may be seen at very low fields since the free carrier density is lower and may become comparable to the trap concentration. However, at a typical driving voltage of OLEDs, trap-free transport is more like an intrinsic behavior.

The mobility of charge carriers, in amorphous organic materials, is strongly electric field and temperature dependent. Electron transport in amorphous organic films takes place by hopping in the LUMO of each molecule. The energy of LUMO can be assumed to have a Gaussian distribution, with a width of the order of 0.1 eV. A packet of carriers propagating in such a system can quickly reach thermal quasi equilibrium at room temperature. Very limited information about the temperature dependence of organic light emitting diode material mobility has been provided in the temperature range from 10 to 70k. In field effect transistors, the mobility in α-sexithiophene (α-6T) displays a nonmonotonic temperature dependence. Above 50 K, the transport is thermally activated, whereas below 40K, the field-effect mobility is approximately temperature independent, as shown in Figure 77.

The hole mobility is also found to increase quasilinearly with gate voltage at room temperature, and that dependence becomes superlinear at low temperatures, as shown in Figure 78. In silicon field effect transistors, more scattering events can lower the mobility a bit when the carrier density is increased by raising the gate voltage. This abnormal behavior might indicate some transport coherence in a high carrier density if grain boundaries and defects in thin films are negligible.

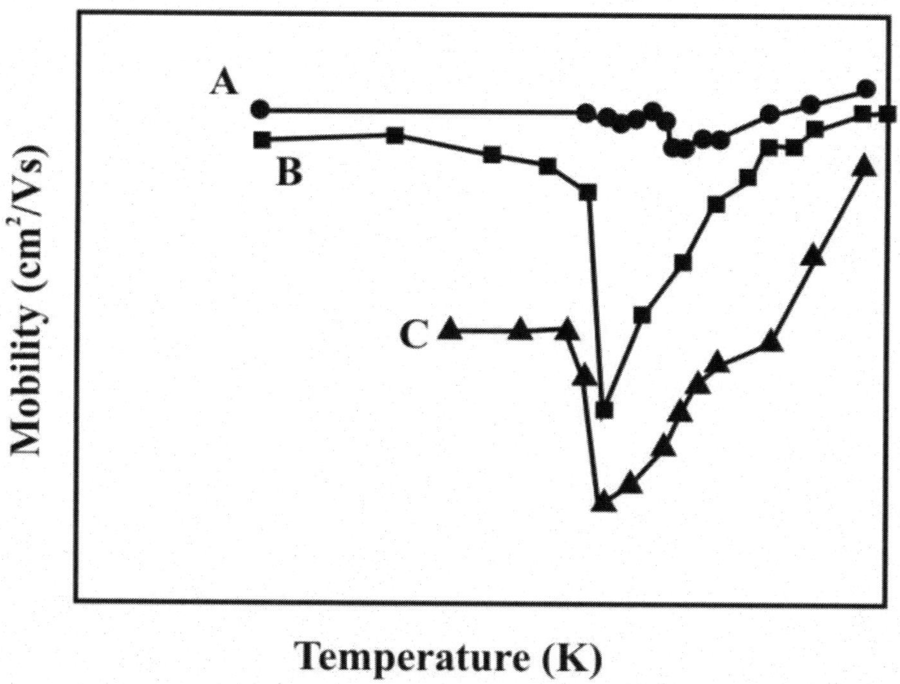

Figure 77. Measured μ vs T for three α-6T TFT' A, B and C.

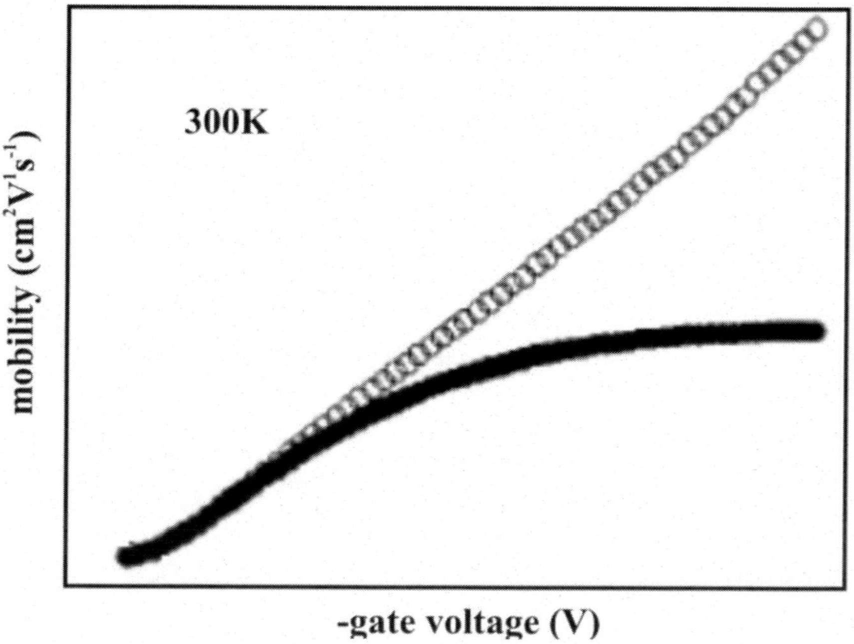

Figure 78. Variation of the hole mobility of a 6T poly-crystalline film as a function of gate bias. Data were recorded at 300K. Closed circles correspond to uncorrected data, and open circles to data corrected for the contact series resistance.

The picture of charge transport is not very clear. However, a lot of effective experimental work aimed at better electron transport has been done. At driving voltages for a luminance of over 1, 000 cd/m^2, there is usually a substantial hole leakage current in ETL, resulting in low current efficiency. N-type doping such as Li can improve the conductivity of ETL and keep a better balance of hole and electron current.

Spin ½ electron and spin ½ hole can form 4 spin combinations, 3 triplets and 1 singlet. If singlet and triplet capture cross-sections are equal, it implies only 25% efficiency for generation of singlet excitons. This factor is assumed to be insensitive to both devices and materials. Since the ground state of fluorescent molecules is typically singlet, the transition from triplet to singlet is usually forbidden by spin conservation. Thus, the maximum internal quantum efficiency is expected to be 25% for fluorescent organic light emitting diode.

Recent studies showed that this is not the case in conjugated polymer. Singlet to triplet ratio depends on the conjugated length. Recombination is spin independent for the monomer, but that spin-dependent process favoring singlet formation is effective in the polymer as a consequence of the exchange interaction, which will operate on overlapping electron and hole wave functions on the same polymer chain at their capture radius. Making use of the triplet excitons can greatly improve quantum efficiency. One approach is to introduce species that will allow efficient triplet luminescence (phosphorescence). This can be provided by high-atomic-number elements with strong spin-orbit coupling. A platinum-containing porphyrin (PtOEP) has been successfully used as a dopant in molecular organic light emitting diode. The internal quantum efficiency can be as high as 87%.

Under typical working conditions, a potential of order 5V might be applied across a 100 nm thick device. This corresponds to an average applied field of 5×10^7V/m. The local electric field is strongly affected by the presence of space charge and the precise details of the device structure. It may be significantly smaller than this value over the recombination zone. In organic light emitting diode, hopping is the dominant carrier transport mechanism so that the mean free path of charge carriers is roughly the molecular separation. This is about one order of magnitude shorter than the range of the Coulomb interaction. Thus at low fields, recombination can be treated using the carrier motion-controlled Langevin approximation. Nevertheless, at a high-applied field, the size of the carrier capture surface will be greatly reduced. The simple Langevin model is no longer valid. This behavior can be explained by the Thomson-like recombination controlled by the carrier capture.

The coupling of electronic excitations to photon states is affected by the physical structure of organic light emitting diode. The presence of the metallic cathode provides a mirror that modifies the pattern of the electromagnetic modes near the cathode. The variation of Alq film thickness in a two-layer structure will change the distance between the emissive zone and the cathode, therefore emission spectra, as shown in Figure 80, and spectral emission patterns will change accordingly. The emission pattern could deviate from the ideal Lambertian pattern. This dependence of emission pattern on Alq thickness cannot be explained by classical optics.

Due to the refractive index mismatch between the emitting layer and air, a large fraction of the light is trapped in the glass, ITO, and organic layers. The external coupling efficiency is estimated by classical theory to be 1/2n2 for large n. The external coupling efficiency in planar organic light-emitting devices can be modeled based on a quantum mechanical microcavity theory and measured by examining both the far field emission pattern and the edge emission of light trapped in the glass substrate. The external coupling efficiency is dependent upon the thickness of the ITO layer and the refractive index of the substrate. The coupling efficiency ranges from 24% to 52%, but in general it is much larger than the 18.9% expected from classical ray optics. It is clear that the results obtained from ray optics can overestimate the internal quantum efficiency.

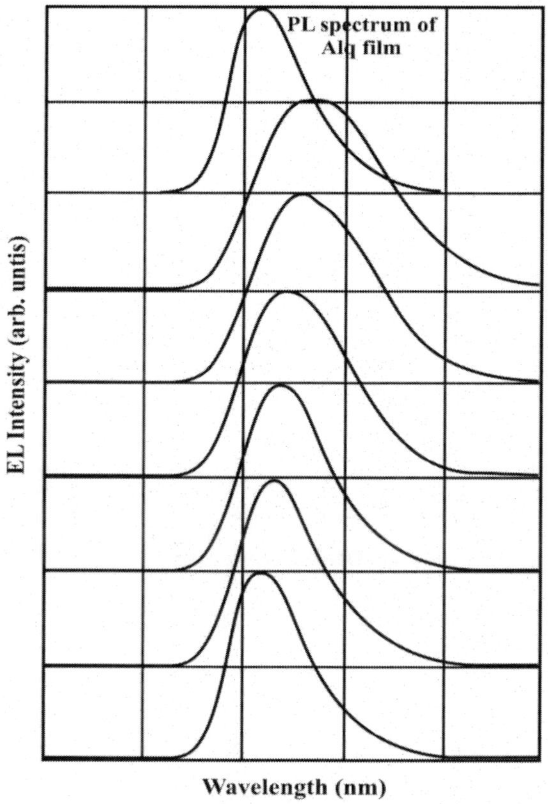

Figure 79. Electroluminescence spectra at normal directions in ITO/TPD (50 nm)/Alq/MgAg devices.

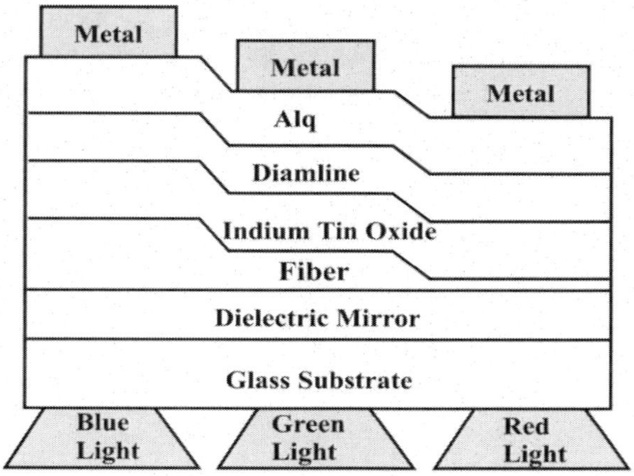

Figure 80. Schematic layer structure of a patterned planar microcavity in which the Si_3N_4 filler layer is etched to three different thicknesses to change the optical properties.

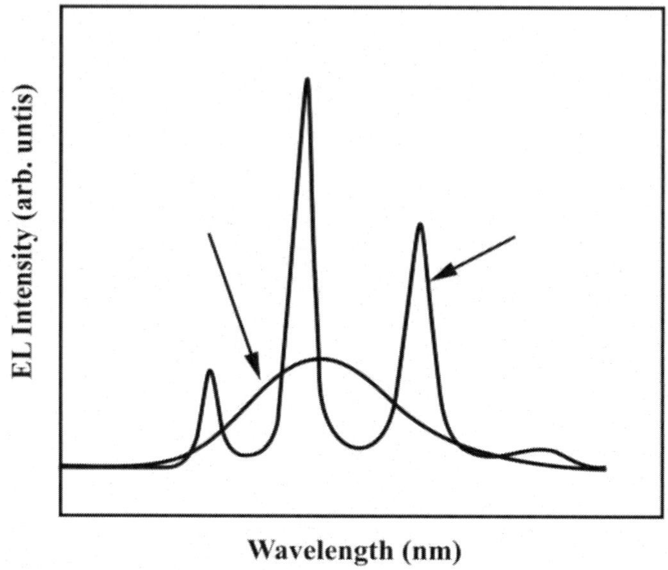

Figure 81. Electroluminescence spectrums from a three-mode microcavity LED, in which the three peaks are at 488, 543, and 610 nm.

After adding a multilayer dielectric stack on the other side of the ITO from the organic layer, this stack, along with the metal cathode, forms a resonant cavity with a narrow band pass that can be selected by varying the optical thickness of the cavity. With an emitting layer that has a white spectrum, the three primary colors can be obtained by patterning different dielectric thicknesses as indicated in Figures 80 and Figure 81. This technology is promising for RGB colors even though close control of the etching process is required.

Here, the detail presented above is the fundamental theory of organic light emitting diodes. Up to the present time, the novel concept of organic light emitting diode is by development through printed electronics in order to obtain the flexible display technology.

Chapter 5

MATERIALS FOR ORGANIC LIGHT EMITTING DIODE AND FABRICATION TECHNIQUE

5.1. MATERIALS FOR ORGANIC LIGHT EMITTING DIODE

5.1.1. Development of Transparent Electrode

After the development of cellulose-based composite as a substrate for display technology, the feasibility of composite for being a substrate for organic light emitting diode has been extensively investigated. From the fundamental point of view, it was remarkable to note that most electro-optic device such as organic light emitting diode require electrically conductive and transparent electrodes. The example of transparent electrode was due to ITO [255-258]. In general, it was commonly deposited on glass substrate. It has been widely used as transparent conducting electrodes in many optical devices because it possesses attractive properties with respect to visible transparency and electrical conductivity. However, the use of ITO as transparent electrode was still limited for mass production in industrial commercialization. It is expensive and the use of indium was considered a rare material. Moreover, the process for ITO requires high temperature in order to obtain good quality. On the other hand, the use of ITO was considered as ceramic and it resulted in brittleness and was subsequently not appropriated for flexible display technology.

One way to overcome this issue was the use of conductive polymer. From the fundamental point of view, conductive polymers were discovered two decades ago. Up to the present time, over 20 conductive polymers have been extensively researched [259-262]. The example of conductive polymer involved polypyrrole, polyaniline, polythiophene and poly (3,4-ethylenedioxythiophene), etc. The use of conductive polymer was still versatile, depending on application. Polypyrrole and polyaniline were commonly employed as halogenated or strong polarity solvent while PEDOT: PSS employed water as solvent. The use of conductive polymers was therefore versatile. From the fundamental point of view, the ability of conductive polymer considered the conduction of charge, which referred to electron moves within and between the polymer chains. The conductivity of polymer stemmed from a combined number of factors such as the combination between single and double bonds, which is commonly called conjugated backbone. In conjugated backbone, single and double bonds contained a less strongly localized π bond. The p-orbital in the π bond overlap each other, allowing the electrons to be more easily delocalized and move more freely between atoms.

In addition, it was important to note that to use conductive polymer with higher efficiency, effect of solvent should be remarkably considered. From the structural point of view, structure of polymer was considered as organic material and the ability of organic materials was very sensitive under solvent. Polymer solution was theoretically considered as "like dissolve like", which referred to polar polymer dissolved in polar solvent while non-polar polymer was dissolved in non-polar solvent, suggesting that the effect of solvent should be carefully considered in order to tailor the excellent properties of conductive polymer. This was to prevent the effect of collapsing chain of polymer if non-appropriate solvents were applied. In case of polymer solution, organic solvent will generate positive charge, or protons and the mobility of protons along a double bond of conjugate polymer can be referred to the ability of conductivity.

The example of the use of conductive polymer for transparent electrode in organic light emitting diode considered the use of PEDOT: PSS. The use of PEDOT: PSS was encouraged in the area of transparent electrode due to feasibility of fabrication by printed electronic, which gained interest for mass production in industrial commercialization. Therefore, the development of transparent electrode from conductive polymer has been extensively evaluated.

Printed electronic is a commonly-used deposition technique for direct patterning or row-column pixel assemblies. It is rapid and simple, with lower operating cost than any vapor deposition technique. In addition, it offers a non-contact patterning feature. This technique has been applied in a variety of materials including ceramics, metals, polymers and composite. A standard desktop computer and a standard inkjet printer can control the delivery of a small volume of ink for precise patterns. Inkjet printing technique has attracted interests of many academia and industries for the commercial market of electronic devices such as organic light emitting diodes (OLED), thin film transistors, solar cells, etc. To effectively apply the inkjet printing technique, parameters such as substrate properties, process parameters and ink solution properties have to be optimized. In general, inkjet printers often require conductive polymer size below 200 nm in order to prevent the clogging of print-head nozzles and to avoid the agglomeration among particles, as suggested by inkjet printer suppliers.

Further, ink formulation is also a key parameter. Several literatures propose conductive inks dedicated to the inkjet printing process by using conductive polymer such as PEDOT:PSS (poly (3,4-ethylenedioxythiophene) poly(styrenesulfonate)). This conductive polymer is commonly used in optoelectronic device due to its solution processability, high transparency in the visible wavelength and excellent thermal stability. PEDOT: PSS is a promising candidate to replace indium tin oxide (ITO) as a transparent electrode in optoelectronic device. There is an imminent need to find new transparent conductive materials to replace ITO, since indium is a scarce material on earth, and the market for optoelectronic devices has been rapidly expanding. In addition, as PEDOT: PSS has high mechanical flexibility; it could be used in flexible electronic devices that are expected to be the next-generation of electronic devices. However, PEDOT: PSS has a problem of low conductivity. The film obtained from the neat PEDOT: PSS aqueous solution usually has remarkably lower conductivity than ITO. Therefore, it is important to significantly enhance the conductivity of PEDOT: PSS. According to several literatures, the conductivity enhancement can be achieved by adding small amounts of solvent to the ink formulation. The ink solution properties such as viscosity and surface tension also have to be controlled by appropriate solvents and surfactants.

Nano-scale size of PEDOT: PSS particle in aqueous solution was observed by TEM. Figure 82 exhibits the particles in irregular shape with good homogeneity in aqueous solution. After filtering, the particle size is in the range of ~ 100-200 nm as the nozzle size of the cartridge print head is 200 nm in diameter. In general, the wide particle size distribution was preferred to fulfill the free space between each particle. Higher packing efficiency of particles with different sizes offers higher electrical conductivity. Higher packing efficiency will also help to avoid the problem of crack phenomena in electronic device fabrication.

Before using PEDOT: PSS in printed electronic for transparent electrode, many parameters have been modified; for example; contact angle, viscosity, surface tension and conductivity of PEDOT: PSS. Deionized water was added in order to dilute the conductive polymer solution as well as to adjust the viscosity of the ink for inkjet printing. On the other hand, glycerol was added to improve electrical conductivity. To add more organic solvent could make the PEDOT chain shift from a coil structure to an expanded-coil or a linear structure; based on electron spin resonance (ESR) microscopy and electrochemical experiments. Such conformational changes may lead to a change in the morphology of the PEDOT: PSS film. The free ion generated from organic solvent can be trapped with polymer chain, so called "polyelectrolyte," enhancing the mobility of free ion along the polymer main chain. However, it is important to note that only the appropriate amount of organic solvent can be added for conductivity enhancement. Excessive solvent can lead to network structure formation among linear polymer chain and reduce the overall conductivity.

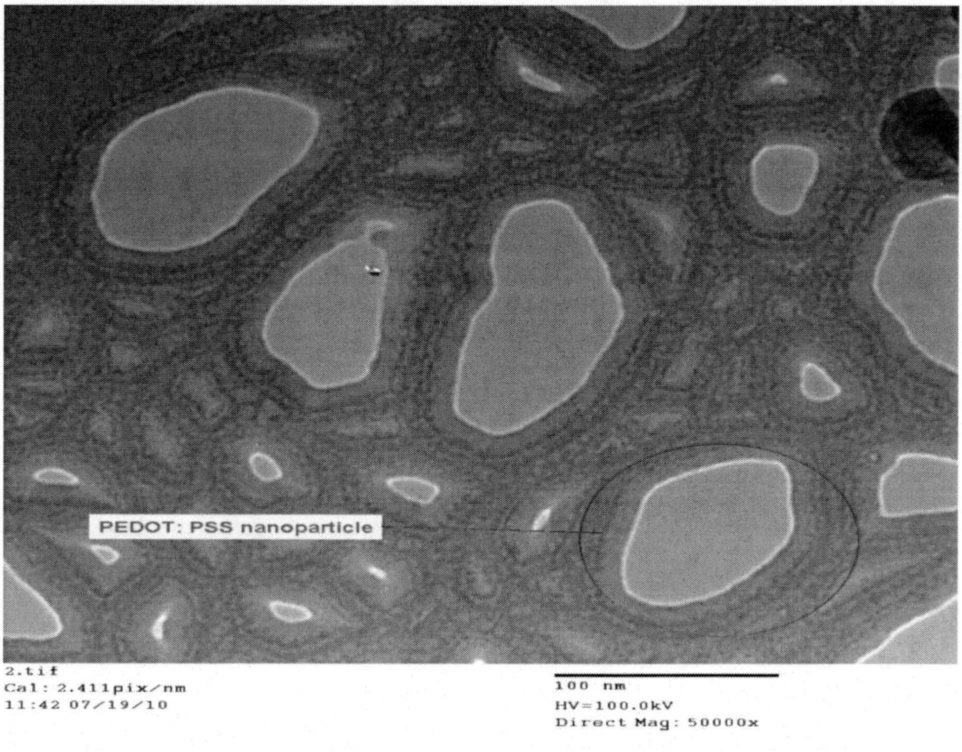

Figure 82. TEM image of PEDOT: PSS nano-particle in aqueous solution [263].

Table 5. numbers of layer and film thickness measured by profilometer

Number of layer	Thickness (micron)
1	1.45
3	
5	4.94
10	8.23
	15.23

The surface tension of PEDOT: PSS decreased from 71 to 48 mN/m after adding surfactant, defoamer and conductivity-enhancer. In general, adding a low concentration of surfactant is useful where micelle formation or micro-emulsion dominates. In this experiment, PEDOT was preferentially dissolved in the micelle assembly because of its hydrophobic nature. Adding the surfactant led to PEDOT particle connection as a consequence of the decrease in surface energy and interfacial tension between PEDOT: PSS solution and substrate.

The decrease in the surface tension of PEDOT: PSS was supported by the decrease in its contact angle. In this experiment, a small amount of liquid was dropped on a flat horizontal solid glass surface. The shape of the droplet is determined by the Young's relation. The contact angle of PEDOT: PSS in aqueous solution at ambient temperature decreased from 44 to 38°. The liquid is strongly attracted to the solid surface, and completely spread on the solid surface.

Table 5 exhibits the film thickness as a function of printing layers. The film was successfully deposited by inkjet printer. It is a general practice that the film thickness is controlled; it must be as thin as possible to prevent the accumulation of electric charge mobility at the interface layer. As traditional electronic devices are fabricated on rigid material, the preferred nano-scale thickness can be achieved to avoid accumulation of electric charge mobility. Our fabrication was expected to find application in flexible display devices. Therefore, the thickness of the anode layer was increased from nano to micro scale range to maintain the device lifetime and to avoid crack phenomena when the device bends.

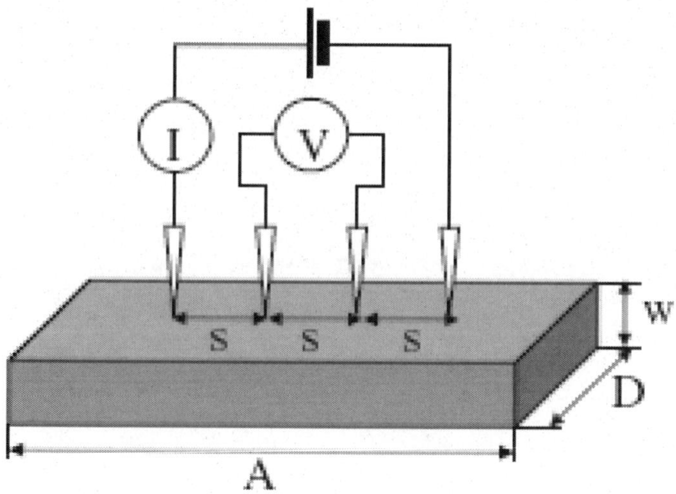

Figure 83. Schematic diagram of test circuit for measuring bar specimen resistivity with the four-point probe method [263].

Table 6. The relationship of numbers of layer and resistivity and conductivity

Numbers of layer	3	5	10
Resistivity (Ω-cm)	25	0.76	0.07
Conductivity (S/cm)	0.04	1.31	14.89

Furthermore, electrical conductivity samples with different thickness were determined by four-point probe method. The schematic diagram of the apparatus is exhibited in Figure 83. By passing a current through two outer probes and measuring the voltage through the inner probes, the substrate resistivity can be determined.

Conductivity, denoted by σ, was determined from sheet resistance data according to the definition of resistivity (ρ).

$$\text{Conductivity} = 1/\,(\text{Resistivity})$$

Table 6 illustrates the electrical resistivity and conductivity of samples printed with 3-10 layers of PEDOT: PSS. Electrical conductivity of PEDOT: PSS anode layer depends on a number of factors including conformation structure of polymer chain, content and distribution as well as the processing method and conditions. However, the most significant parameter determining the properties of PEDOT: PSS anode is the number of deposited layers. As a rule, the conductivity of film increases with increasing the number of deposited layers. In our experiment, in case of one deposited layer, the resistivity was beyond the capability of the four-point probe apparatus as the deposition was not sufficient. Three, five and ten deposited layers successfully measured the electrical properties.

Figure 84 exhibits the number of layer and conductivity relationship. In the case of organic light emitting diode devices, the optimization between the small amount of number of layers and conductivity must be considered in order to preserve high transparency. In OLEDs theory, light is generated after the combination of positive and negative charge at emissive layer. The direction of light is passed though the transparent anode. According to Figure 87, the photograph exhibited that 3 and 5 deposited layers still yielded transparent properties. In the case of 1 deposited layer, the transparency is clear, although its electrical conductivity is too low to be detected. As for 10 deposited layers, the electrical conductivity is high, but the transparency became too low to be useful for OLEDs application.

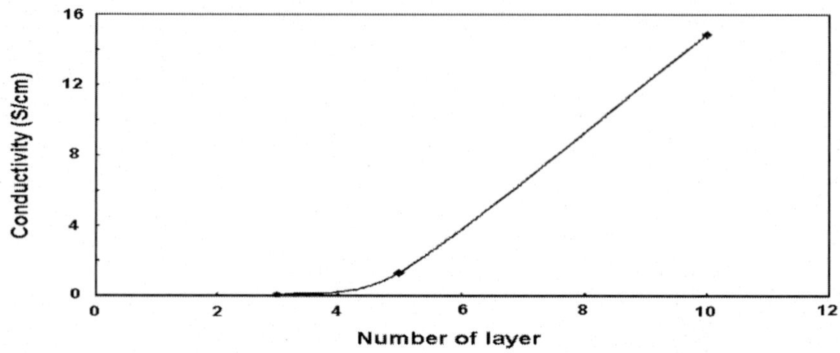

Figure 84. Linear relationships between number of layers and conductivity value [263].

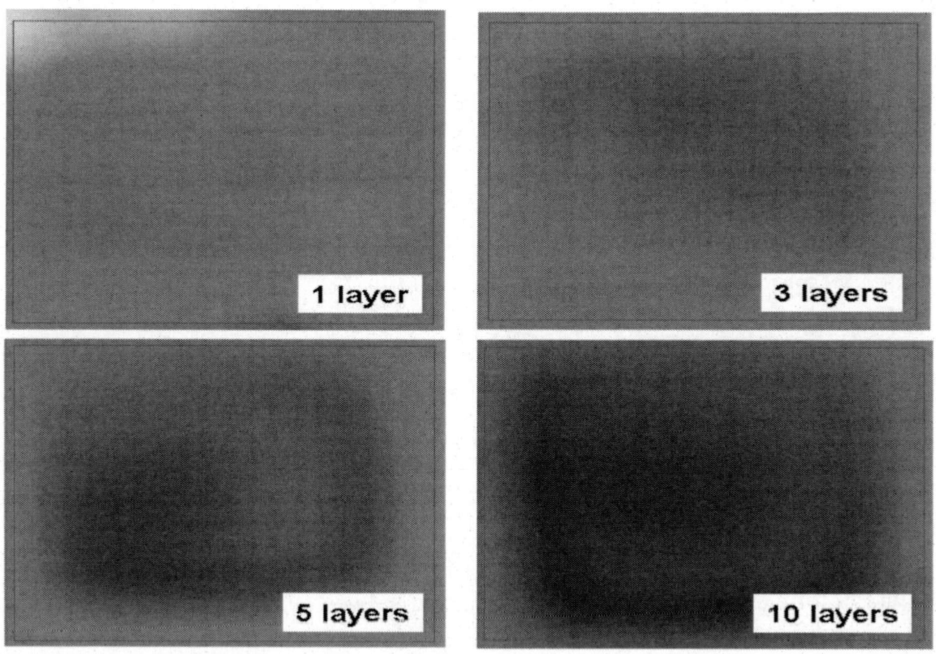

Figure 85. Photograph of PEDOT: PSS as anode by Desktop inkjet printer [263].

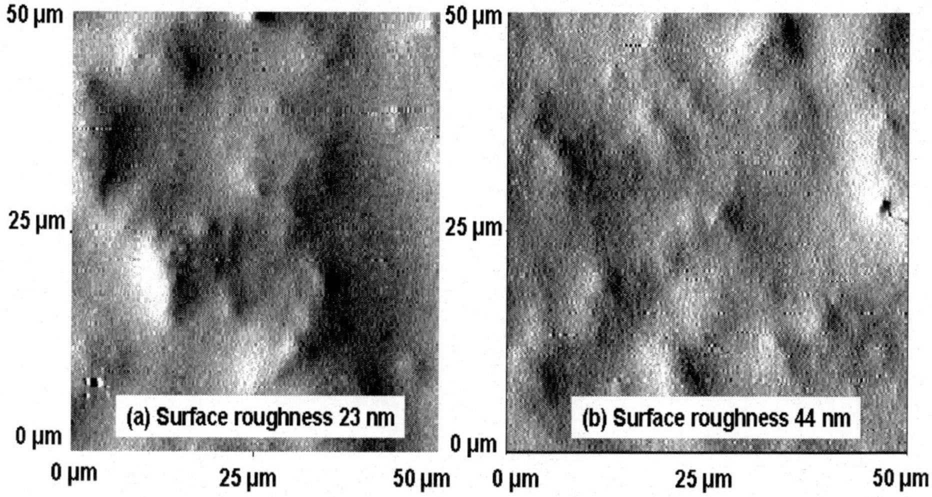

Figure 86. AFM investigation of PEDOT: PSS coated on resin in (a) 3 and (b) 5 layers, respectively [263].

Figure 86 illustrates AFM image taken with lateral contact mode. AFM image with scan size of 50x50 μm^2 was acquired for the surface morphological observation. In Figure 88, three and five layers of deposited PEDOT: PSS film was investigated. The root mean square (rms) surface roughness values of these 3 and 5 layers are 23 and 44 nm, respectively, indicating that the smooth surface of PEDOT: PSS can be deposited on cellulose based composite substrate in flexible display technology.

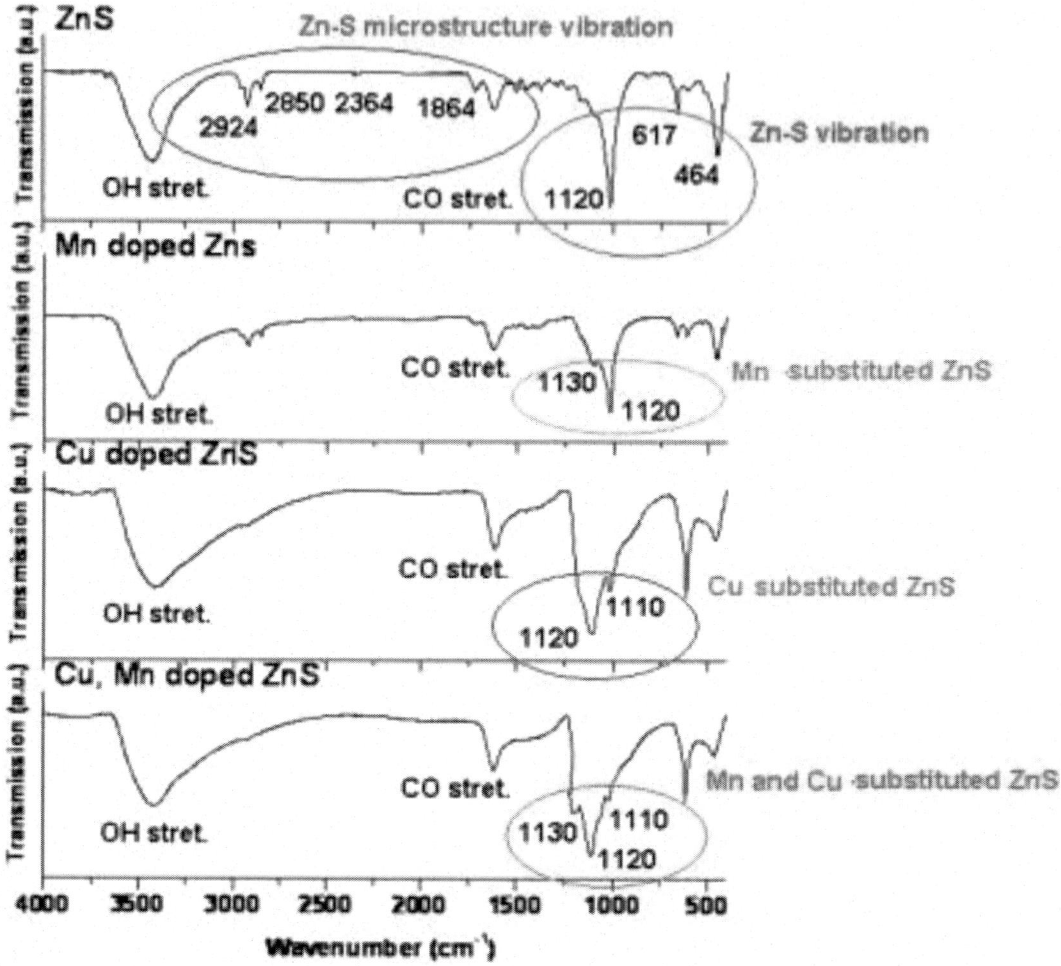

Figure 87. FTIR spectra of ZnS and any metal-substituted into ZnS powder as emissive layer [267].

5.1.2. Development of Emissive Layer

After development of the transparent electrode, the operation of organic light emitting diodes on the emissive layer have been considered. To control the color of light, brightness and efficiency, their development on emissive layer has been evaluated. From the fundamental point of view, the emissive layer was considered as the source of light that can be activated from the combination of both positive and negative charges in this layer. Moreover, the development of emissive layer for optical devices was realized on ceramic-based materials. It was related to metal oxide, which was considered a semiconductor material. To obtain light from emissive layer, activation can be therefore enrolled. The example of metal oxide for emissive layer was considered on ZnS based material [264-266].

As a wide band-gap semiconductor with a range of band-gap energy of 3.6e3.9 eV, ZnS has been applied to make numerous optical devices such as ultraviolet light-emitting diodes, flat panel displays and thin film electroluminescence. Various ZnS crystal structures including bulk and nanoparticles such as spheres, rods, tubes and wires have been successfully synthesized using diverse methods and their luminescent properties have been investigated. In recent years, accordingly, Mn-doped, Cu-doped and Mn-Cu-codoped ZnS powders have received much attention in research because many functions can be added by transporting and controlling numerous types of spin state. In addition, it can be indicated that the properties are closely related to the concentration of metal-doped because it can change the energy band and form luminescence of different energy levels. Many approaches have been used for the preparation of pure ZnS and metal-doped ZnS and control of their morphology has been reported. In 2001, Jiang et al. have exhibited the in situ formation method for ZnS nanowires in the liquid crystal template. In 2005, Charinpanitkul has synthesized ZnS nanoparticles in microemulsion and investigated the effect of surfactant on particle size. In 2009, Li et al. and Salavati-Niasari et al. prepared ZnS via layer-by-layer self-assembly technique and hydrothermal synthesis, respectively. Recently, in 2010, Uekawa et al. also found that ZnS nanoparticles can be successfully prepared by heating the mixture of ZnS precipitate and ethylene glycol. However, these synthetic methods are relatively more complicated. The high reaction temperature in operation process and the use of organic solvent should be preferably avoided. In addition, the fluorescence intensity is far below the minimum requirement of optical devices at room temperature. The development of a novel and simple synthetic route has therefore been considered in order obtaining high yield of product. Figure 87 exhibits the FTIR spectra of ZnS and any metal-substituted into ZnS powder as emissive layer.

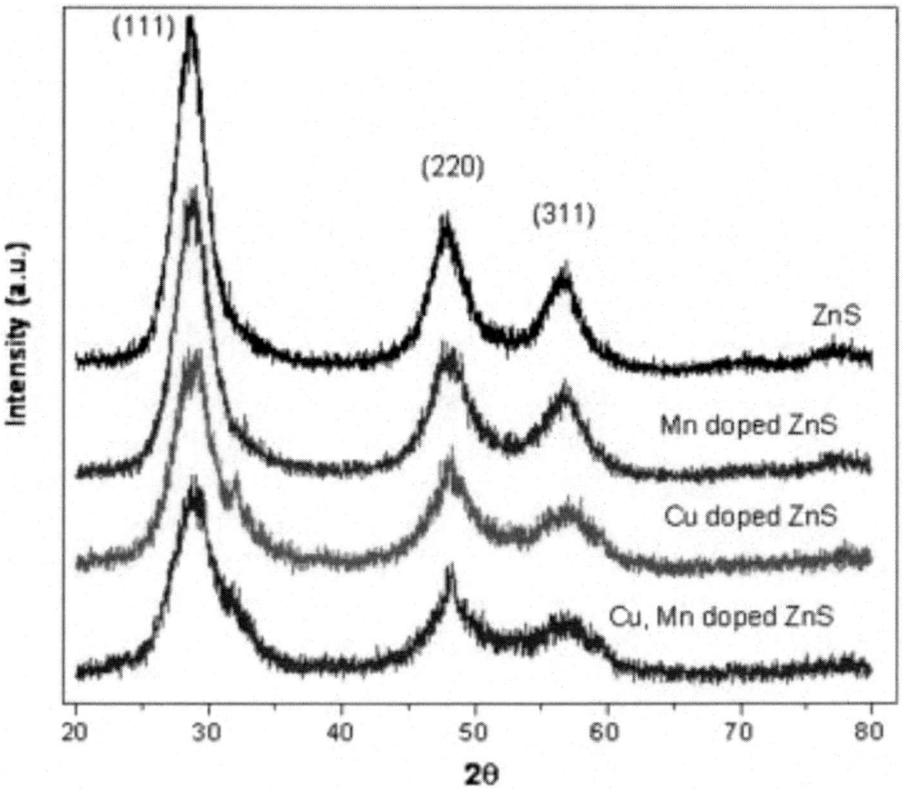

Figure 88. XRD pattern of ZnS and any metal-substituted into ZnS powder as emissive layer [267].

FTIR spectra of ZnS and ZnS doped with Cu and Mn are showed in Figure 89. The characteristic ZnS vibration peaks can be observed at 1120, 617, and 464 cm^{-1}. The peaks at 2924, 2850, 2364 and 1624 cm^{-1} are due to microstructure formation of the samples. The broad absorption peak in the range of 3000-3600 cm^{-1} corresponding to -OH group indicates the existence of water absorbed in the surface of nanocrystals. The bands at 1500-1650 cm^{-1} are due to the C-O stretching modes arising from the absorption of atmospheric CO_2 on the surface of the nanocrystals. The FTIR spectrum of ZnS doped with Mn shows peaks similar to the spectrum of pure ZnS particle. The peak at 1120 cm^{-1} split into two peaks, i.e., at 1120 and 1130 cm^{-1}, indicating that the doped Mn affected the structure portion of the ZnS particles. In parallel, Cu doping had also an effect on the structure of ZnS. The peak at 1120 cm^{-1} split into two peaks, i.e., at 1120 and 1110 cm^{-1}, respectively. It can be explained that either Mn or Cu atom can be partially substituted into Zn position in ZnS crystal. In addition, it can be observed that Mn and Cu- codoped were successfully carried out. The peak at 1120 cm^{-1} split into three peaks, i.e., at 1110, 1120 and 1130 cm^{-1}, suggesting that both Mn and Cu atom were successfully substituted into the main ZnS crystal. The peaks appearing at 1110 and 1130 corresponded with ZnS-Mn and ZnS-Cu, respectively.

XRD pattern of ZnS and ZnS doped with Cu and Mn are shown in Figure 88. Three well-defined diffraction peaks corresponding to the lattice planes of (111), (220) and (311) can be observed in all samples. These peaks matched very well with the cubic zinc blended structure (JCPDS No. 05-0566), confirming the purity of the synthesized ZnS. As ZnS was doped with Mn and Cu, the (111) peak became broadened in all cases, along with the smaller relative intensity of peak (111), suggesting that the crystallinity of ZnS was degraded as a result of doping. Because of the size effect, XRD peaks broaden as the particle becomes smaller. However, in the case of Cu-doped samples, the XRD patterns showed additional peaks at 32. The incomplete substitution occurred during the experimental reaction step. Partial amounts of copper cannot be well substituted due to non-sufficient energy. Broadening of peaks indicates the nanocrystalline nature of the sample from XRD patterns. According to the FWHM of the most intense peak (111) plane, the average crystallite sizes of ZnS and metal-doped ZnS were therefore evaluated. The crystallite size of ZnS was 30 Å, but it dropped slightly to the range of 22-28 Å as it was doped with Mn and Cu. The lattice parameter of ZnS also decreased. This is due to a very small amount of impurity, and these ions were doped into ZnS lattice.

Scanning electron microscope and energy dispersion analysis in Figure 89 exhibits the typical microstructure FESEM image of as-synthesized undoped ZnS and metal-doped ZnS. The powders exhibited blocky particles with irregular shapes possibly due to the agglomeration among particles. Therefore, the actual size of the nanoparticle cannot be determined from the FESEM images as it is limited by its agglomeration and the resolution of the FESEM instrument used. On the other hand, an X-ray energy dispersive spectroscopy (EDX) spectrum (the inset in Figure 79) acquired from the metal-doped ZnS confirms that the particles consist of Zn, S and related metal-dope (Mn and Cu) with a mainly stoichiometric ZnS composition. The successive substitution of Mn and Cu-doped into ZnS cubic structure was successfully prepared. The amount of Mn and Cu atoms was indicated on the EDX spectrum, which is consistent with the added-amount in the preparation step.

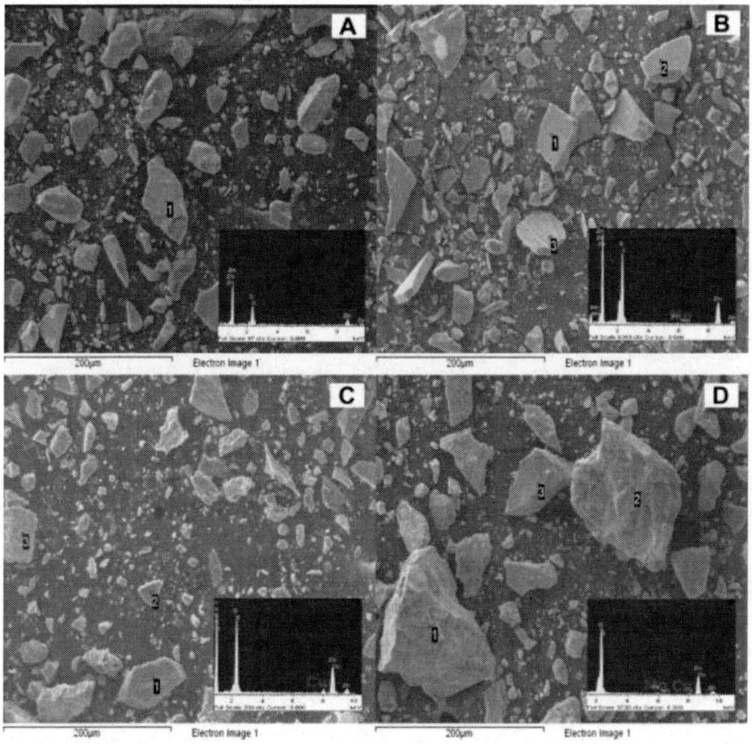

Figure 89. SEM image and elemental analysis of ZnS and any metal-substituted into ZnS powder as emissive layer [267].

Furthermore, X-ray fluorescence was used to determine the quantitative analysis of all samples. The major elements of all samples are Zn and S, while the minor elements are Mn and Cu. In case of Mn-doped ZnS and Cu-doped ZnS, the existence of Mn and Cu is 3.141% and 6.481%, respectively. However, in case of Mn-Cu-codoped ZnS, the existence of Mn and Cu is 7.904% and 2.330%. This result can confirm the presence of Mn-doped ZnS, Cu-doped ZnS and Mn-Cu-codoped ZnS. This analysis method is accurate with an experimental error of 10%. This is due to the counting statistic during measurement.

The fluorescence and room temperature photoluminescence (PL) spectra of ZnS and metal-doped ZnS are illustrated in Figure 90.

ZnS particle and metal-doped ZnS particles were respectively dispersed on clean glass slides for fluorescence microscope observation. It shows a fluorescence image of ZnS and metal-doped ZnS particles without excitation, where various ZnS and metal-doped ZnS clusters were aggregated by small nanoparticles. In case of pure ZnS particle (A), under a UV light of 320 nm excited, the ZnS particles emit strong blue fluorescence light (Right hand side). The photoluminescence spectrum in Figure 90A exhibits a sharp PL at 450 nm, completely coinciding with the blue fluorescence light. On the other hand, for the Mn-doped ZnS, a super-bright yellow with bluish fluorescence was observed when excited with 450 nm wavelength. In parallel, for the Cu-doped ZnS, a super bright green fluorescence was observed at similar excitation as the previous experiment. It exhibited a sharp strong PL band located at 595 and 517 nm, respectively. It can be noted that the strong yellow and green fluorescence are attributed to the transition emission from the introduced energy level of manganese and copper atoms. Thus, metal-doped ions, as optically active luminescence centers, efficiently created the luminescence of particles. After the metal-doped step, the luminescent centers were transferred to impurity ions. For the Mn-Cu-codoped ZnS, Mn^{2+} and Cu^{+} were then substituted into Zn^{2+} sites. The transition metallic ions have different transitions, which resulted in the yellow emission. However, because of starting chemical reagents, Mn^{2+} and Cu^{+} were derived from $MnSO_4$-H_2O and CuCl, respectively, suggesting that the matching oxidation number between Mn^{2+} and Zn^{2+}. Mn is more completely substituted in the Zn position than the Cu atom. However, in this Mn-Cu-codoped ZnS, Cu was also used as substituted metal. It is controversial that XRF and EDX qualitative analysis exhibited Cu atom. Due to the mismatching oxidation number of Cu and Zn, Cu cannot be well substituted into ZnS crystal and may exist as another impurity (not Mn-Cu-codoped ZnS). The green color cannot be detected. On the other hand, the PL band located at 457 and 595 nm show both blue and yellow regions, suggesting partial substitution of Mn atom into ZnS. The sample may contain Mn-Cu-codoped ZnS and neat ZnS. It can be explained that Cu atom may affect this codope substitution step. Consequently, the substitution ability of Mn in Mn-Cu-codoped ZnS may be decreased from Mn-doped ZnS. However, this issue will be discussed later on.

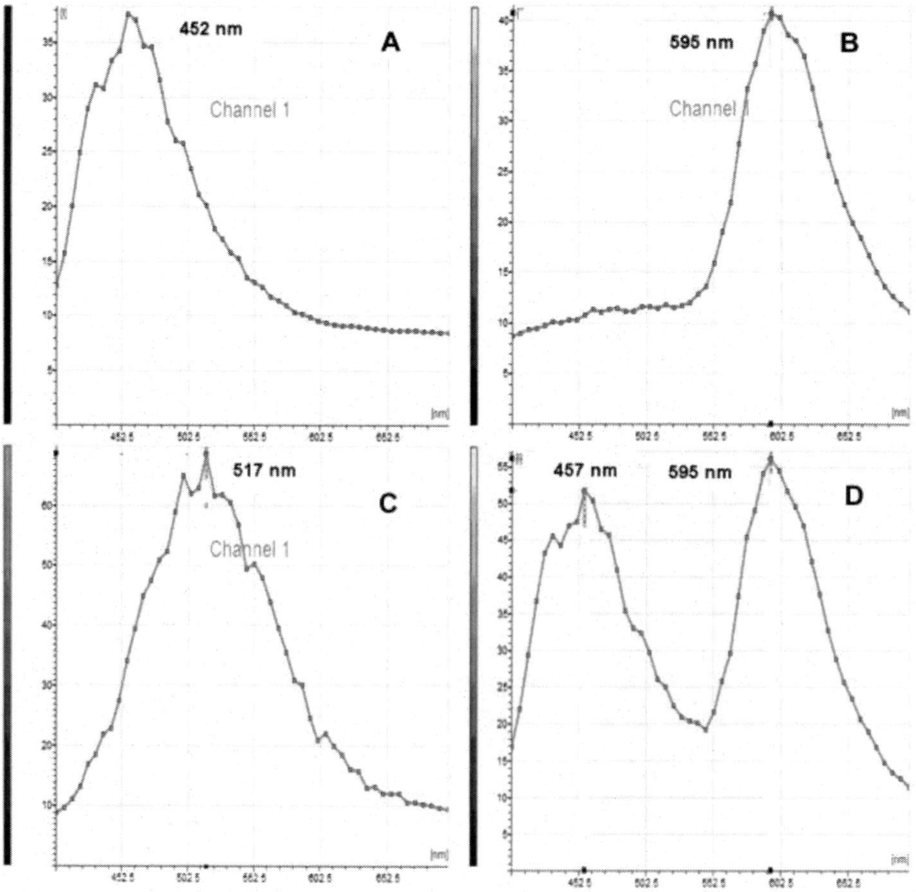

Figure 90. Fluorescence and room temperature photoluminescence (PL) spectra of ZnS and metal-doped ZnS [267].

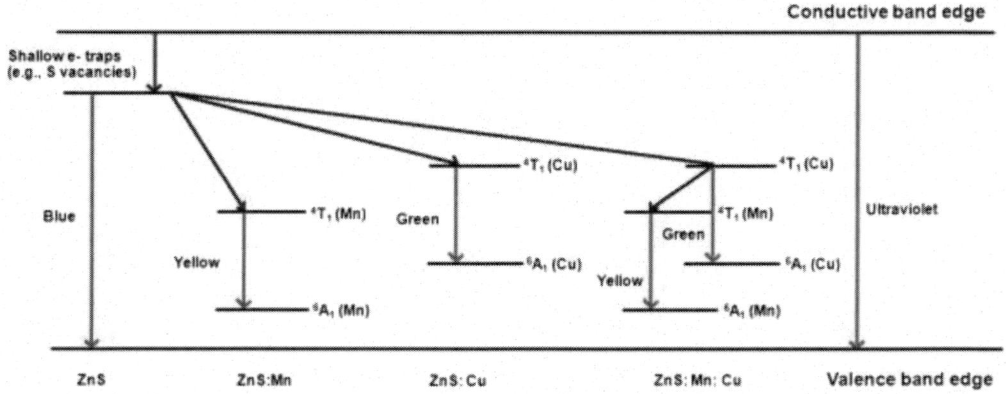

Figure 91. band energy absorption of ZnS [267].

On the other hand, in theory, it is well-known that metal-doped ZnS can be considered a p-type semiconductor. Mn and Cu state related emission is always expected at a lower energy position compared to the band energy absorption as shown in Figure 91. From the fundamental point of view, a wide variety of semiconductors can be expected to generate tunable dopant emissions in different spectral windows once they are successfully doped with Mn and Cu. Either Mn or Cu substitute Zn ions in ZnS crystals introducing a trap energy level, where electron and hole can be trapped. An electron can undergo a photo-excitation process in the host ZnS lattice of nanoparticles and consequently decay via a non-radiative transition from the 4 T1 level to the 6 A1 level. The strong emission could be attributed to the radiative decay between those localized states of Mn and Cu inside the ZnS band-gap. This particular type of material can be utilized as fluorescent material in organic light emitting diode displays.

Furthermore, although ZnS was successfully investigated on its feasibility as emissive layer, the use of ZnS was still limited in flexible display technology. The way to overcome this concern involved the integration of small amounts of ZnS to the polymer matrix as composite, called the hybrid organic-inorganic emissive layer. The advantage of the hybrid layer caused the excellent correlation between emissive layer and its flexibility.

Advances in research are dependent on the possibility of having different properties in the same material, ranging from an optimized processing step, excellent micro- or nano-scale structure, along with the functionalities required for the specific desired application. In recent years, many groups of researchers have successfully prepared nanocomposite systems by incorporating inorganic nanoparticles into an organic matrix. These hybrid nanocomposites provide a path for a novel class of materials, which inherits properties of both the inorganic nanoparticles and the organic matrix in order to tailor excellence in chemical and physical properties. These include mechanical strength, chemical resistance, thermal stability, as well as optical properties.

In particular, a typical hybrid material was contained as an organic phase, bound with inorganic nanoparticles. Polyvinylpyrrolidone (PVP) is a class of organic compounds that has a well-defined structure with an N-vinylpyrrolidone monomer connected as a long chain. The incorporation of inorganic nanoparticles into a polymer matrix can result in significant improvements in a variety of chemical and physical properties. On the other hand, in the case of inorganic materials, ZnS nanoparticles have been widely used as transparent electrodes in solar cells, thin film transistors, as well as a blue light emitter. The most important characteristic of this nanoparticle is the exciton binding energy of ZnS. Chen et al. reported that the exciton binding energy of ZnS is 39 meV. Due to this energy band gap, the exciton is stable at ambient temperature even in a bulk particle. Owing to these properties, ZnS is an excellent candidate for optoelectronic devices such as light emitting diodes. In our previous work, ZnS and related metal (Mn, Cu)-substituted ZnS were successfully synthesized by a wet chemical synthetic route at room temperature. The reaction of zinc acetate and sodium sulfide was conducted in water. The advantages of this synthetic route are a high yield of product and relatively low energy costs. Nevertheless, ZnS and the related metal (Mn, Cu)-substituted ZnS colloid exhibited agglomeration among particles due to the existence of water. Bulk particles of ZnS and metal (Mn, Cu)-substituted ZnS can generate PL intensity, however, the use of these particles is limited due to the device fabrication process. The use of the inkjet printing technique, a commercially available method of fabricating light emitting devices, has been widely employed in the market. The use of printed nanoparticles has the limitation of particle size less than 200 nm, suggested by the supplier. To quantify the ZnS nanoparticle growth (less than 200 nm), size measurements of quantum dots are necessary. A potential protective material is the organic matrix, which includes polyvinylpyrrolidone (PVP) in order to control the effective size and size distribution, but maintaining the photoluminescent properties.

ZnS nanoparticles were successfully synthesized by the wet chemical synthetic route. The ZnS solution exhibited a milky-white color. The viscosity, measured by Ubbelohde type canon, was 14 cgs. It was thus acceptable for use in desktop inkjet printers. The range of viscosity should be in 10–20 cgs at room temperature, suggested by the inkjet printer supplier. In addition, this technique of synthesis effectively offered a high yield of the product and excellent colloidal stability. In this synthetic procedure, polyvinylpyrrolidone (PVP) was chosen as the stabilizer for forming the precursor of ZnS. It can offer the ZnS nanoparticles exposure to the carboxylic functionality of PVP to the outer environment, leading to a very hydrophilic surface, which makes the ZnS stable in water. However, the limitation of as-synthesized ZnS exhibited agglomeration among the nanoparticles; preferable use of this particle must be less than 200 nm. In order to prevent this phenomenon, PVP is used to control and stabilize the particle size of ZnS. Therefore, we developed an analogous procedure for making the as-synthesized ZnS as a hybrid composite film. Figure 92 exhibits the FTIR spectra of ZnS/ PVP hybrid composite.

The absorption FTIR spectra of the ZnS/PVP solution and ZnS/PVP hybrid composite film were recorded in the range of 4000 to 500 cm−1 and are exhibited in Figure 92. In the case of the ZnS/PVP hybrid composite film, the peaks appearing at 1200 and 620 cm^{-1} are due to Zn–S vibration and at 2920, 2360 and 1640 cm^{-1} are due to microstructure formation of the sample. On the other hand, this is in contrast to the ZnS/PVP solution, the peak appearing from 1200 to 500 cm^{-1} was featureless. This is due to the small amount of ZnS nanoparticles in solution. However, the broad absorption peak in the range of 3410–3465 cm^{-1} corresponding to –OH group indicated the existence of water absorbed in the surface of the nanoparticles. The presence of this band can be clearly attributed to the adsorption of the same atmospheric water during measurement. The bands at 1600 and 2370 cm^{-1} are due to the C–O stretching modes arising from the absorption of atmospheric CO_2 on the surface. The XRD patterns of the neat PVP film, ZnS embedded PVP film and bulk ZnS are exhibited in Figure 93.

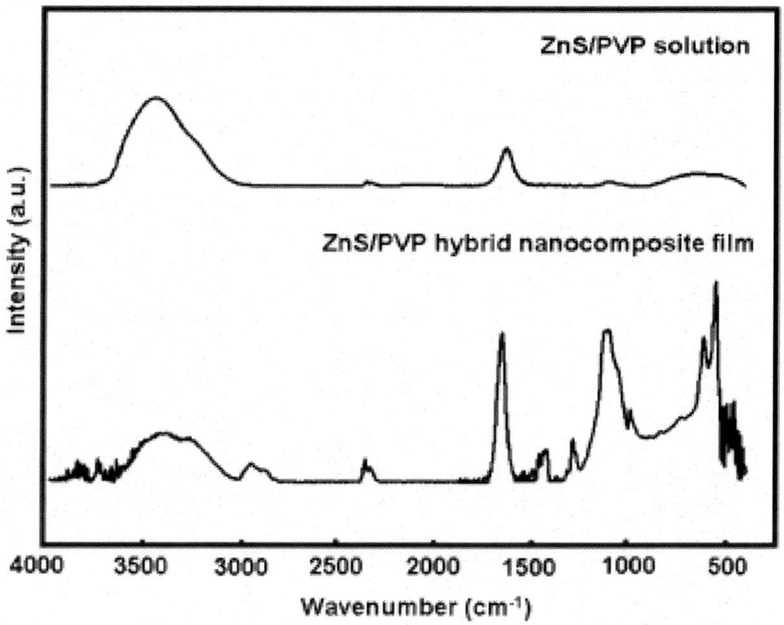

Figure 92. FTIR spectra of ZnS/ PVP hybrid composite [268].

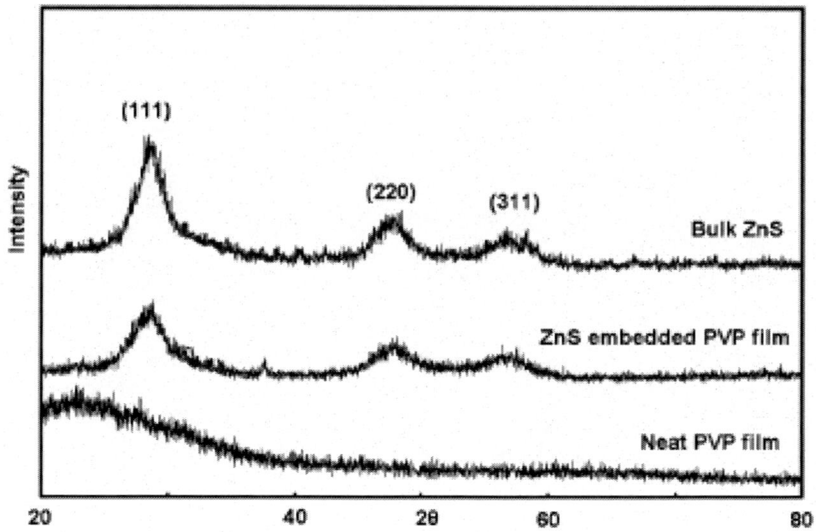

Figure 93. XRD patterns of the neat PVP film, ZnS embedded PVP film and bulk ZnS [268].

An X-ray diffraction technique was used to investigate the crystalline structure of film materials. No diffraction peak was detected for the neat PVP film, suggesting that it did not contain any crystallite structure. However, more diffraction peaks appeared in the case of the ZnS embedded PVP film. Three positions, remarkably corresponded to the lattice planes of (111), (220) and (311), are very well matched with the cubic ZnS structure (JCPDS No. 05-0566). This result is consistent with bulk ZnS.

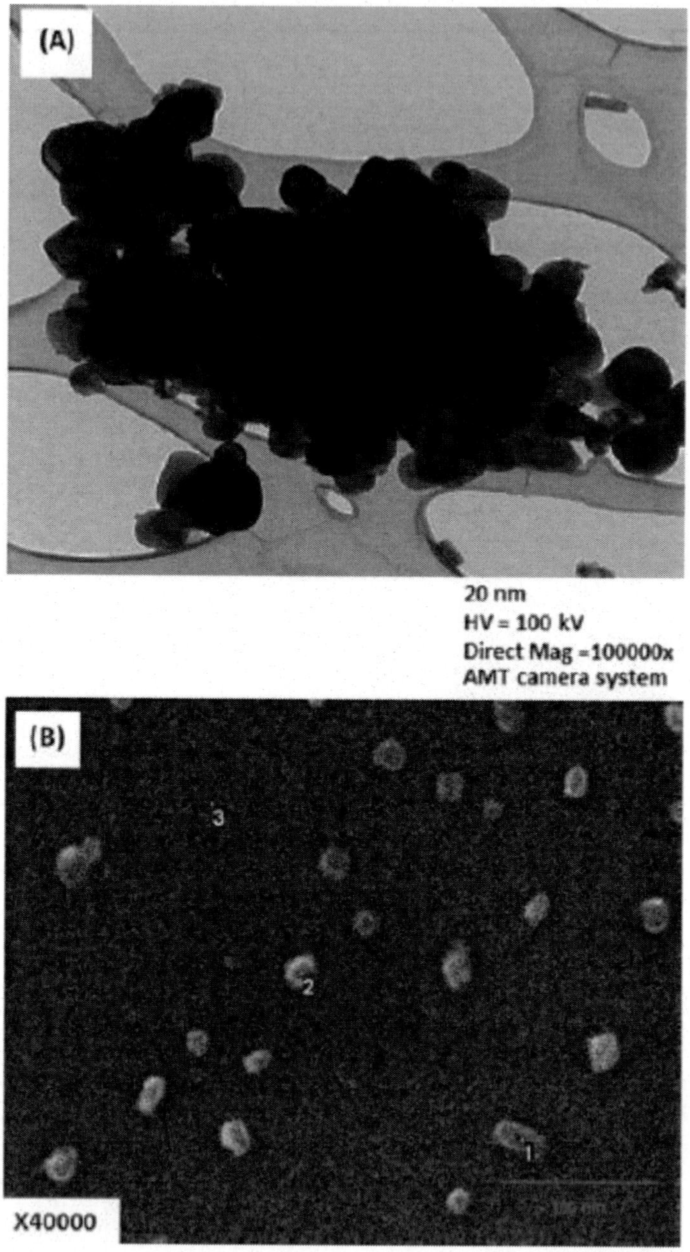

Figure 94. TEM and SEM analysis of ZnS/PVP hybrid composite [268].

The TEM and SEM micrographs of the ZnS nanoparticles are given in Figure 94. The particle size was estimated with the distribution histogram that exhibited the maximum frequency for the states particle range. The shape of the nanoparticles exhibited spherical form, and the size was about 30–50 Å. A comparison of particle sizes obtained XRD and TEM illustration in good agreement. On the other hand, the SEM image exhibited an agglomeration character among nanoparticles due to solvent removal. Three points on the SEM image show the position of EDX analysis. This revealed that the amount of zinc and sulfur atoms can be detected. Figure 83 shows that ZnS embedded PVP was successfully deposited on our substrate. The SEM image revealed that the morphology of the deposited film has a uniform surface. The inkjet printer offered a printed film with high density on the printed area. No space can be observed. The ZnS nanoparticles had successfully penetrated the surrounding PVP matrix. The presence of white color represented the position of ZnS nanoparticles. The homogeneous distribution of ZnS was successfully controlled using a sonicator on the ink solution before printing. However, the film still exhibited a crack-like character. Before SEM observation, the hybrid ZnS embedded PVP was deposited on a flexible substrate. The optimization between mechanical properties and expected device efficiency needs to be considered in order to prevent crack phenomena. On the other hand, the insert in Figure 95 exhibits the EDX analysis. It shows that the deposited film was mainly composed of zinc and sulfur atoms within the scan area (pink color). The presence of carbon and oxygen was possibly due to the use of carbon tape and retained the solvent after the deposition step. Figure 95 exhibits SEM and EDX analysis.

After, the printed hybrid organic–inorganic ZnS embedded PVP film was observed by laser confocal microscope. Figure 95 shows a fluorescence image of ZnS nanoparticles without excitation, where numerous ZnS clusters were aggregated from smaller particles. Under excitation of an UV light of 320 nm, they emit strong blue fluorescence light. The blue color corresponds to those clusters.

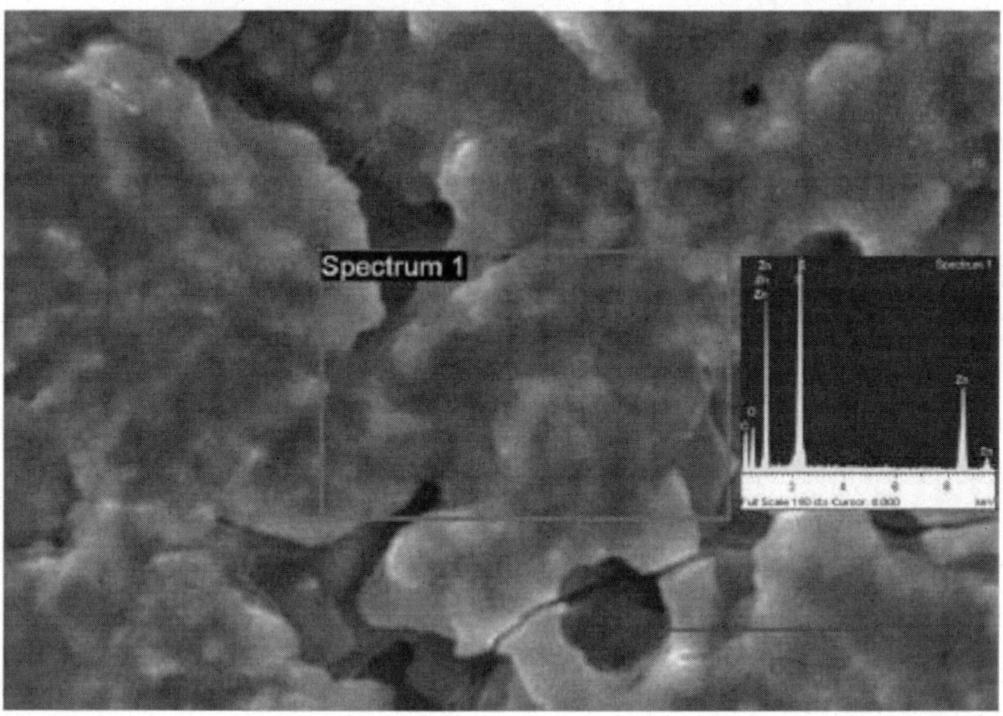

Figure 95. SEM and EDX analysis of ZnS/PVP hybrid composite [268].

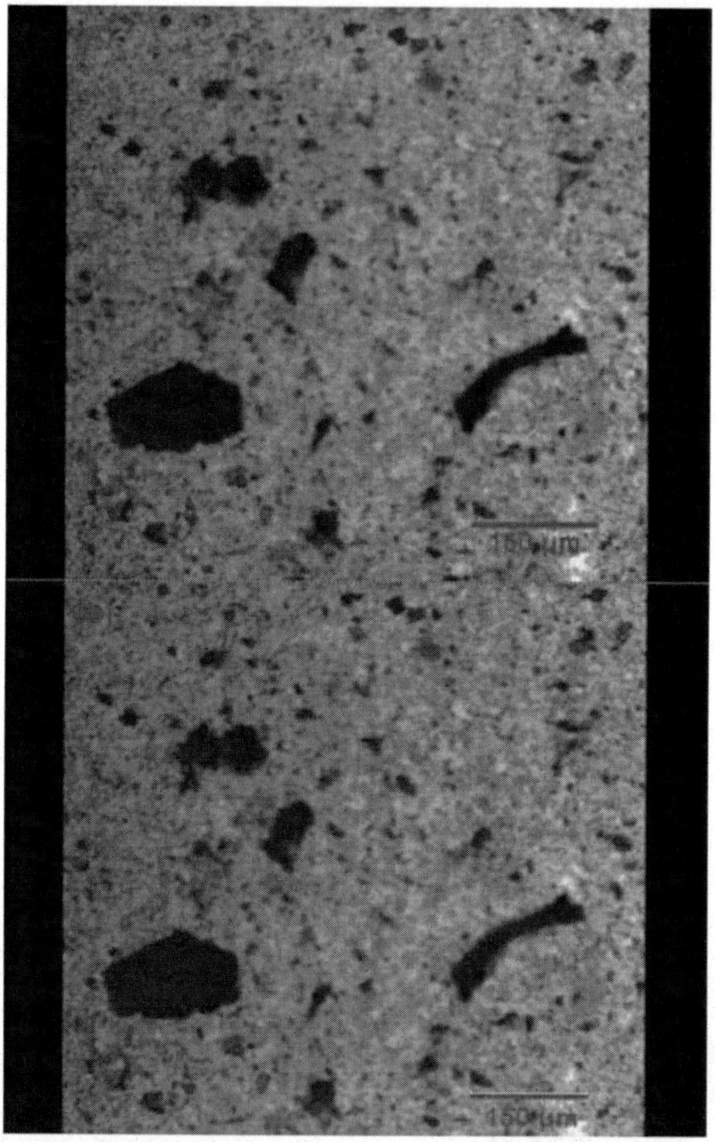

Figure 96. shows a fluorescence image of ZnS nanoparticles with and without excitation [268].

5.1.3. Development of Cathode Layer

To develop a circuit for organic light emitting diode, the last deposition involved cathode formation. A cathode provided the positive charge and it was therefore combined with negative charge from the anode. The combination of both charges occurred in the region of the emissive layer and subsequently activated the light. In order to evaluate the efficiency of the cathode layer, numerous approaches were developed on the feasibility of fabrication.

Recently, attention has been focused on the development of convenience and low-cost processing techniques to deposit conductive features on cellulose based composite substrate for flexible display technology. Direct-write techniques, one type of printable electronics with the feature of low conductive ink consumption has attracted interest in many industrial applications such as printed circuit boards and display devices, because they promised the simplification of device manufacturing. Among multiple direct-write technologies, drop-on demand inkjet printing technology is currently used as a high potential method in the printing industry due to low instrument cost and fewer processing steps compared with other patterning processes. In addition, mark-less and non-contact patterning approaches are also beneficial. This technique has been applied with a variety of nano-scale materials including ceramics, metal, polymers as well as composites. Conductive ink solution, a mixture of metallic particles and organic solution, had been agitated before loading into the inkjet cartridge, in order to prevent agglomeration of particles. The common condition of their particle size must be less than 200 nm and monodispersed. The diameter size of the nozzle of an inkjet cartridge is less than 200 nm, indicated by the printer supplier. In particular, an inkjet printer uses water as media to suspend ink particles. It is desirable that ink particles at the end product should be well aligned and give high electrical conductivity after heat treatment at a relatively low temperature. In the case of metal nanoparticles, when the ink solution was deposited onto the substrate, the solvent was then evaporated with moderate heating and a thin solid film of nanoparticles was obtained. The thin solid film was further heated to a higher temperature in order to remove organic molecules and ensure that nanoparticles sinter into a dense metallic film. In general, nanoparticles are based on monodispersed form. Based on the theoretical maximum density, the attainable packing fraction for the dried film should be 74% if considering a perfect face-centered cubic lattice. However, the nanoparticle can be randomly packed; this fraction is reduced to 62.5%. Higher dispersions and various different sizes of ink nanoparticles offer larger packing ratio production. Figure 97 exhibits the dissociation of silver salt and the chemical process on the use of stabilizer.

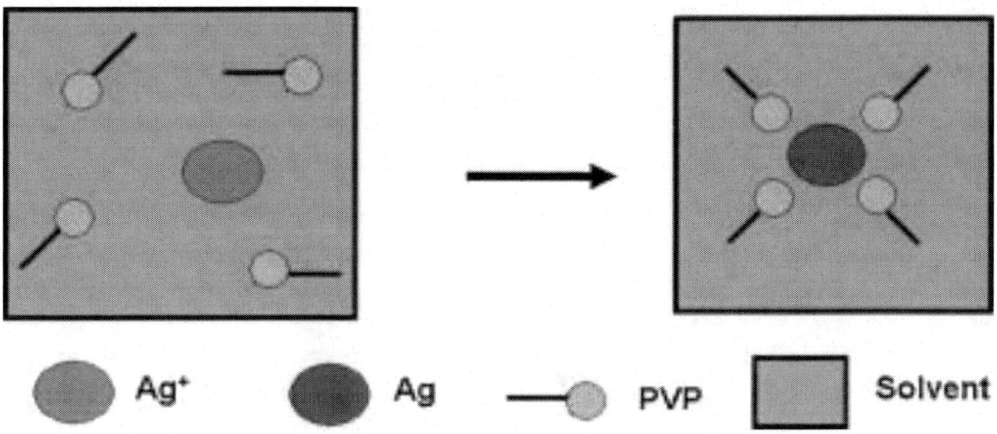

Figure 97. Dissociation of silver salt and the chemical process on the use of stabilizer [269].

In this experiment, silver nanoparticles were successfully synthesized using polymer in wet chemical synthetic route. PVP was used as the stabilizer for nanoscale size-of-particle control. Embedding such particles in the PVP matrix is also advantageous. The formation of silver particles was synthesized from silver ions using a reducing agent. It can be trapped along the PVP chain and the size of this particle can be controlled into nanoscale. It can be explained that PVP can prevent the agglomeration and precipitation among each particle size. The bulk particle of silver can be avoided. On the other hand, in common theory, it is remarkable to note that silver nitrate has the highest lattice energy when compared to any other silver salt and therefore, it can be dissolved only in polar solvents. The silver ion is a powerful oxidant and can be reduced to metal silver. In our case, silver nitrate was dissolved in ethylene glycol in order to synthesize silver nanoparticles in the organic phase. The chemical reaction requires that some of the organic molecules can dissociate lattices of silver nitrate and form organic-soluble complexes as shown in Figure 86. PVP, one type of organic molecule, was used to dissociate silver ions. PVP acts as a protective agent and restricts the mobility of silver ions during the reaction, controlling most of the agglomeration. The in situ produced silver ions were bound to PVP through a chemical bonding, which later helps silver ions to get converted into silver. According to related theory, it can be explained that the coordinative interaction between the organic molecule and metal is related to electron donation from the organic molecule to the metal atom and electron reverse-donation from the metal atom to the organic molecule.

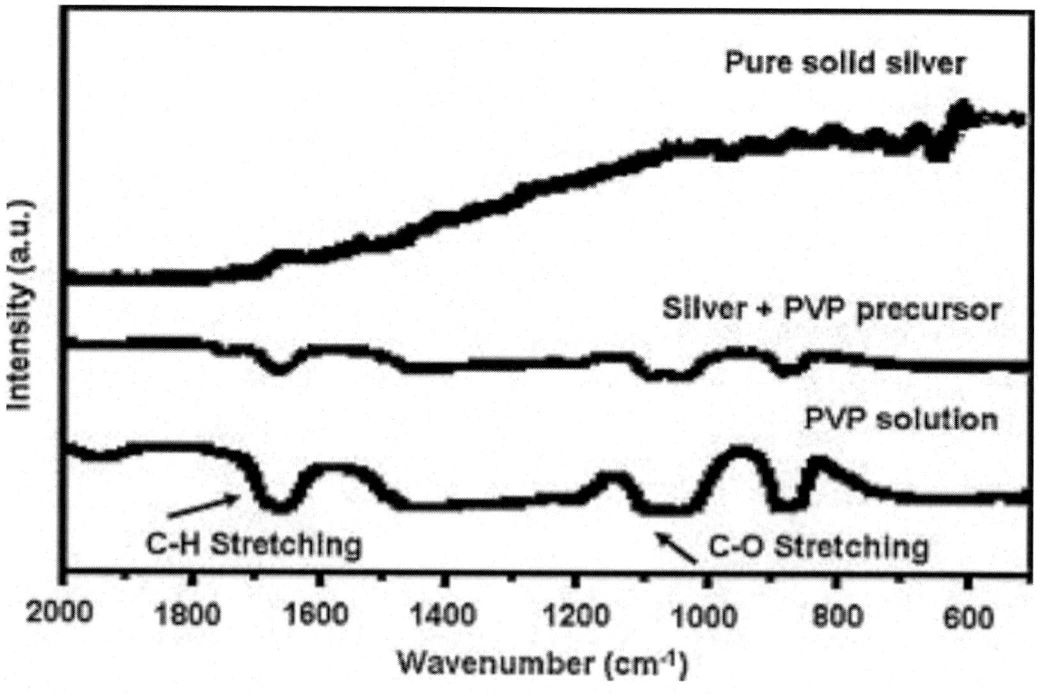

Figure 98. FTIR spectra of as-synthesized silver [269].

FTIR spectroscopy was used to investigate the neat PVP solution, thin solid films of silver and coordinative interaction between silver ions and PVP solution. Figure 98 exhibits FTIR spectra of pure silver, PVP solution and a mixture of silver nanoparticles dispersed in

PVP solution. For the PVP solution, it can be found that the pattern of PVP molecules exhibited absorption peaks at wavenumbers of 1663 and 1180 cm^{-1}. These corresponded to C-C stretching and saturated C–O vibration, respectively. In addition, it can be observed that the intensity pattern of the mixture of pure silver nanoparticles and PVP solution was slightly decreased from pure PVP solution. It can be explained that neat silver nanoparticles, generally exhibited as neat colloidal nanoparticles, may disturb the adsorption on the PVP molecule. On the other hand, for the silver film, the spectra exhibited a featureless character over the identical wavenumbers. It can be interpreted that the organic molecule from PVP solution was completely removed after centrifugation and annealing at 200°C.

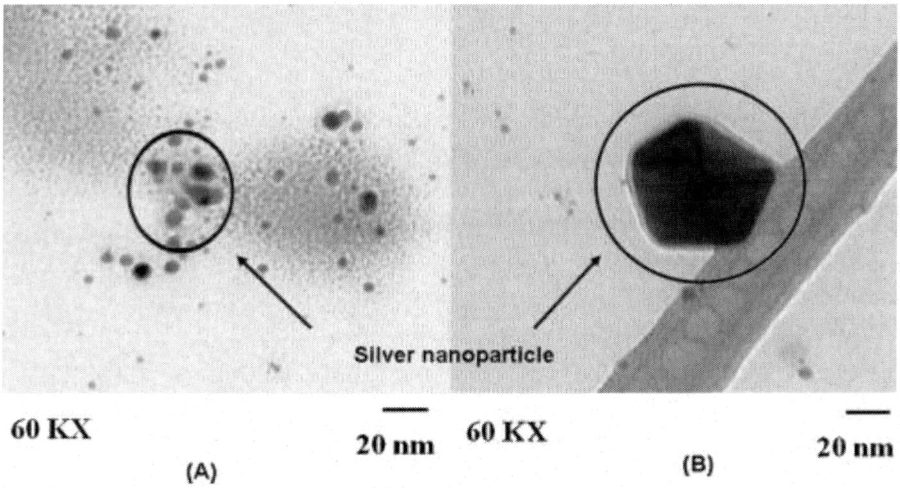

Figure 99. Distribution of silver particle observed by TEM [269].

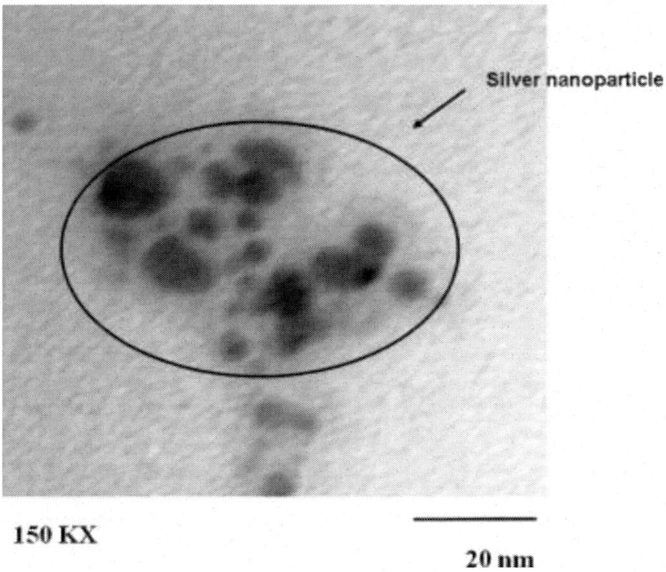

Figure 100. Morphological properties of silver particle observed by TEM [269].

The particle size, microstructure and size distribution of synthetic nanoparticles were characterized via TEM. Figure 88 indicates that a homogeneous distribution of the nanoparticles can be observed. The smaller size of nanoparticles was prepared in PVP (Mw 40,000) with a diameter of roughly 20 nm, while the larger ones were prepared in PVP (Mw 10,000) with a diameter of roughly 100 nm. The results were consistent with XRD observation.

The size distribution of both small and large silver nanoparticles was quite uniform. In particular, a very large size and non-spherical shape particle can be seen at higher magnification in Figure 99. It is important to note that the PVP solution plays an important role in stabilizing the suspension of silver particles and prevents them from agglomerating. XRD was used to investigate the crystalline structure of the materials.

Figure 101 exhibits the XRD patterns of specimens deposited with different silver particle sizes. No silver diffraction peak was detected for the mixture of silver nitrate and PVP solution. The precursor solution did not contain the crystallite structure of silver. More diffraction peaks appeared in the case of silver thin film deposition including large and small silver particle sizes and the mixture of these. The peaks can be remarkably observed at 37.58, 43.78, 63.98 and 76.88 corresponding to the face-centered cubic structure of silver crystal planes (111), (200), (220) and (311) of silver metal, which are consistent with those of bulk silver. It can be determined that silver is well crystallized. In Figure 85, it is evident that the intensity of the diffraction peak increased from a small particle size to a large particle size and the mixture of these created higher density thin film materials. From the Scherrer's equation, the particle size of synthesized silver has been calculated to be roughly 26 nm and 110 nm for the expected small and large particle size based on the (111) peak, respectively. This is close to the value deduced from TEM images.

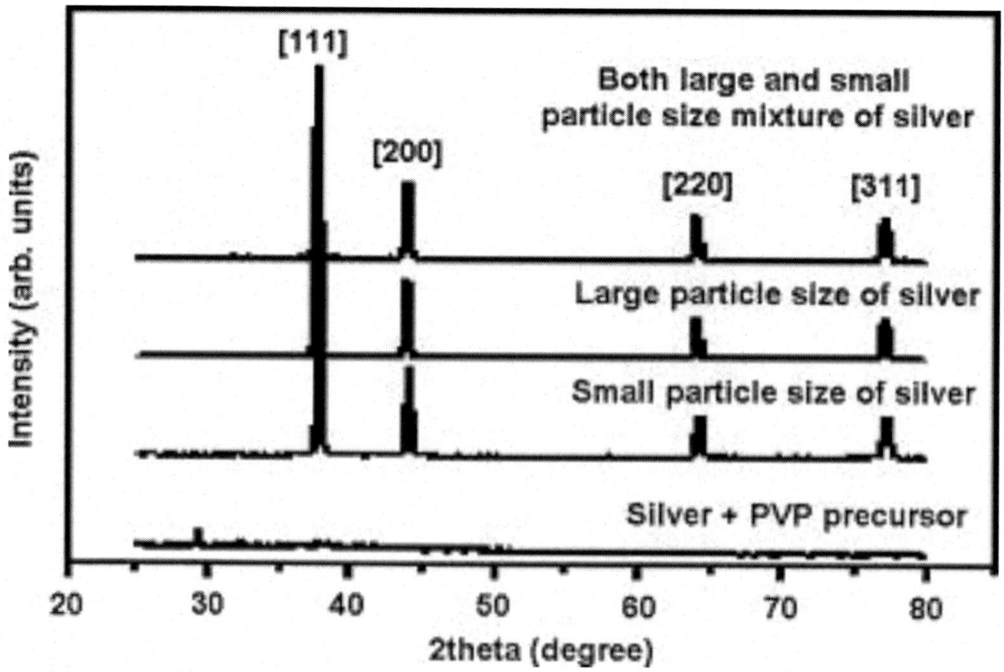

Figure 101. XRD pattern of as-synthesized silver [269].

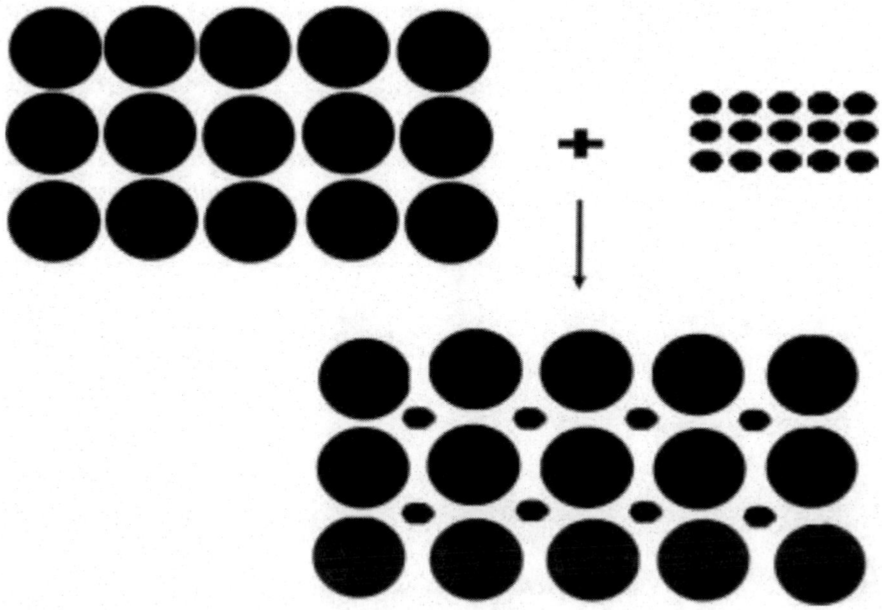

Figure 102. Schematic drawing models for packing efficiency [269].

The packing efficiency of silver nanoparticles and electrical conductivity of both large and small sizes of silver nanoparticles were successfully synthesized using the wet chemical synthetic route. The mixture of both large and small particle sizes was performed in order to enhance the electrical properties and to prevent crack and pinhole phenomena, while a thin solid film of silver was fabricated as the electrode layer of an electronic device. The crack and pinhole phenomena can occur at the grain boundary of the silver film or free spaces among silver particles. The connection between grains of silver is easy to crack due to higher residual stress when silver film is bent. The free space among silver particles has less silver, so when silver film is bent, the non-connection between aligned silver occurred.

Figure 102 represents the schematic drawing model. It was used to predict the two dimensional packing of silver nanoparticles. In order to prevent the occurrence of space among large particles of silver colloid, smaller particles can fill in this space. The combination of both large and small particle sizes of silver colloid was expected to provide a dense thin solid film of silver. After the heating step, both large and small particle sizes of silver were packed. The free space among silver particles in the mixture case is less than the individual ones.

In Figure 103, the dried silver of both large and small particle sizes provided the average diameter as 100 and 20 nm, respectively. However, each procedure of both synthetic steps may offer a significant amount of polydispersity in particle size. The size of silver particles may be higher or lower than 100 and 20 nm. After silver particles were heated to 200°C for solvent evaporation, the free space can then be observed. In case of small particle size (right top), the large space was not well observed, while this is opposite for large particle size (left top). In case of the large and small particle size mixture, the dried silver film exhibited denser packing than film of only large or small particle sizes. The necking area among particles was expected to enhance when silver dried film was heated with higher temperature. Once silver nanoparticles were printed as a thin solid film, the relationship of current density and applied voltage was measured using an I-V probe station. From the measurement in Figure 104, under forward bias voltage from 0–20 V, the current density was exhibited as 0.1 J/cm^2. This constant current density and range of applied voltage are commonly utilized for OLEDs devices. All the measurements were conducted at ambient atmosphere with no encapsulation. The consistent result of our synthesized silver offered data identical data to the commercial silver paint electrode. The preliminary experiment of electrical conductivity was successfully performed for further development of cathode layer in organic light emitting diode in the near future.

Figure 103. Morphological properties of as-synthesized silver observed by SEM [269].

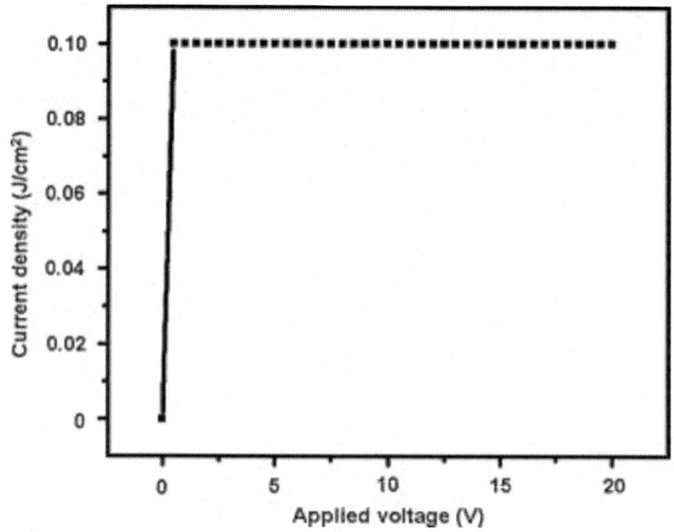

Figure 104. I-V characteristic curve of as-synthesized silver [269].

5.2. Fabrication Technique

Due to the flexibility, ease of use, and low cost of processing, printers are often currently used to develop many types of electronic devices including transistors, diode, energy storage devices as well as sensor-actuators. In the past, printers only focused on the graphic printer, in which text and figures were required and its application commonly referred to the book, journal and magazine industry. This technique not only used small amounts of ink quantity, but the rapid and time-controllable concept was beneficial, which allowed commercialization developments. However, in the electronics community, electronic devices are generally prepared as a stack of thin film, and the printer can be effectively used. The challenge was to design and replace ink solution with an electronic-sensitive feature of ink quality. The rapid and time-controllable processes of the printer, used for electronics, were successfully developed for the electronic community [270-272].

The use of the printing technique to manufacture electronic components constituted an emerging class of materials with potential applications including transistors, diodes, energy storage devices, and electronic memory devices as well as sensors and actuators. This printing technique offered several advantages over laboratory scale and subtractive batch processes, such as photolithography and vacuum processes. Printers for electronic applications generally required more continuous and homogeneous patterns with restrictions on the layer thickness, and roughness, as well as resolution. This was probably due to the minimum requirements on electronic device whose electronic quality depended not only on the printer but also compatibility with electronic substrate.

Moreover, from the structural point of view of electronic components, which generally fabricated devices such as multilayer thin film, the integration of functional ink deposited on substrate had to be considered because printed ink solution does not dissolve the pervious film. However, in order to solve this issue, Ummartyotin et al. studied the effect of viscosity on ink solution. The addition of glycerol in ink solution can not only increase conductivity but also reduce ink solution viscosity. The rate of flow of ink solution was consequently low. On the other hand, Maattanen et al. proposed the use of UV irradiation and surface plasma treatment treated on the surface of substrate. The improvement on surface energy and adhesion force among ink layers was successfully enhanced. This probably led to higher quality performance of electronic devices. After that, we have to evaluate the electronic print quality. It was generally difficult to determine the efficiency of the device during the fabrication process. Before determination of print quality, the compatibility between substrate and ink layer had to be considered. Surface roughness, thickness, line widening, line-edge roughness, surface resistivity, volume resistivity, and electronic performance had to be considered.

Finally, it was very hard to select the appropriate printer for excellent printed electronics components. Printers for electronic purposes typically used common printing equipment or other low cost equipment suitable for defining patterns on substrate. Examples of printer can be generally categorized into different sub-processes such as gravure, flexography, offset, screen and inkjet printing. All sub-techniques have their advantages and disadvantages, and thus modifications of the conventional techniques were extensively required in order to print excellent patterns for electronic applications. The selection of printing techniques for use should be considered based on the requirements of lateral resolution, printed thickness, and printing speed as well as ink properties.

Gravure printing, one of printing processes, has been extensively used for electronic devices such as antennas, capacitors, and organic light emitting diodes as well as solar cells. Generally, gravure equipment comprised a 2-roller system where the coating roller had an engraved pattern. The coating roller was partially inserted into a bath of the ink that was continuously refilled as mentioned in Figure 89. The print patterns were engraved into a metallic roller by a laser or chemical etching. A high quality of print pattern can be consequently achieved. This technique was generally appropriated for large print volume. The advantage of gravure was that low viscosity inks worked very well and very high roller speed was possible 1-10 ms^{-1}, whereas, the disadvantage of this technique concerned surface roughness of substrate, which may interrupt the ink transfer to substrate. In addition, the use of this technique is preferred for low printing pressure. High print pressure limits the use of this technique to applications where there were no previously printed soft layers or high surface roughness.

From an industrial point of view, the gravure quality can be extensively controlled by adjusting the printing speed, pressure, and roughness of surface as well as properties of ink. Depending on device quality and quantity, the gravure technique can be successfully used for continuous printed pattern for electronic applications.

Flexographic printing is generally similar to gravure printing. The difference of flexographic printing from gravure is that the printed image stands up on the printing roller, the surface of which is typically made from rubber. The typical flexographic printer consisted of fountain rollers that continuously transfer ink to the ceramic anilox roller, which had engraved cells/micro cavities embedded into the exterior. This allowed the collection of ink, which transferred to the relief on the printing roller that performed the final transfer to the web. The ink picked from the anilox corresponding to the negative pattern of the motif, as mentioned in Figure 106.

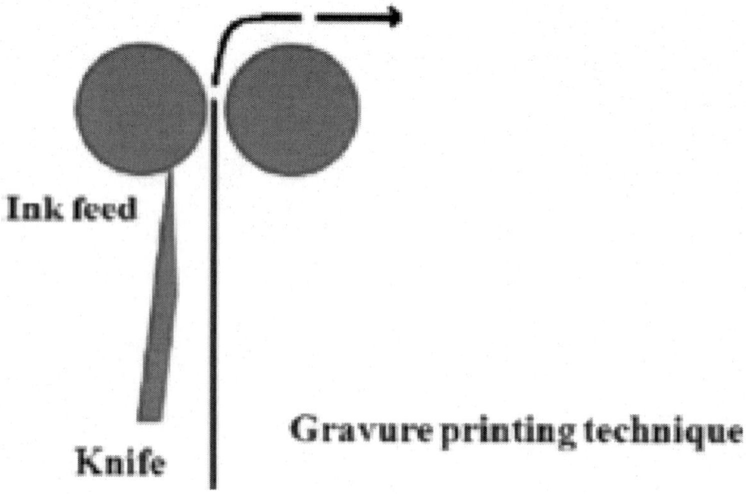

Figure 105. A schematic of gravure printing technique [273].

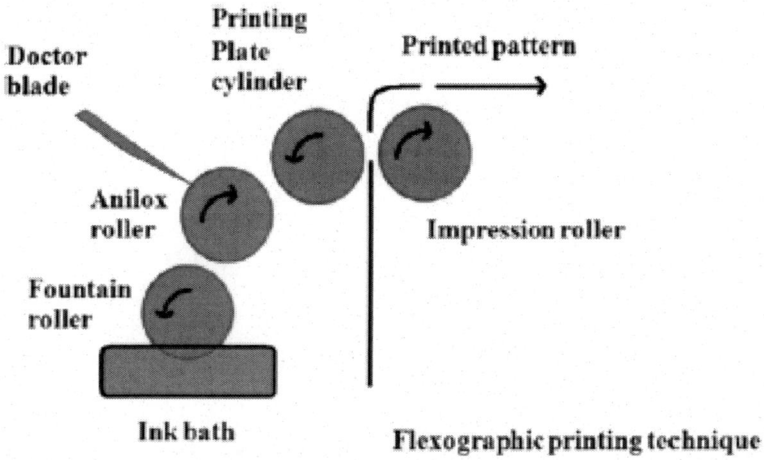

Figure 106. A schematic drawing of flexographic printing technique [273].

Offset printing, one of the most common printing techniques, involves ink solution being transferred (or offset) from a plate to a rubbery blanket then to the printing surface at a rather high pressure. In particular, the ink conditions for this technique generally required high viscosity (5-200 PaS) with a suitable surface tension (30-60 mN m^{-1}). The printed pattern on the offset roll was based on a difference in surface energy, which is generally created by photolithography. Typically, the non-printing areas on the printing roll were moistened with water before the ink was printed on the surface, as exhibited in Figure 106. This technique offered numerous advantages including consistency of image quality for a large area of printing, and low cost of processing and printing.

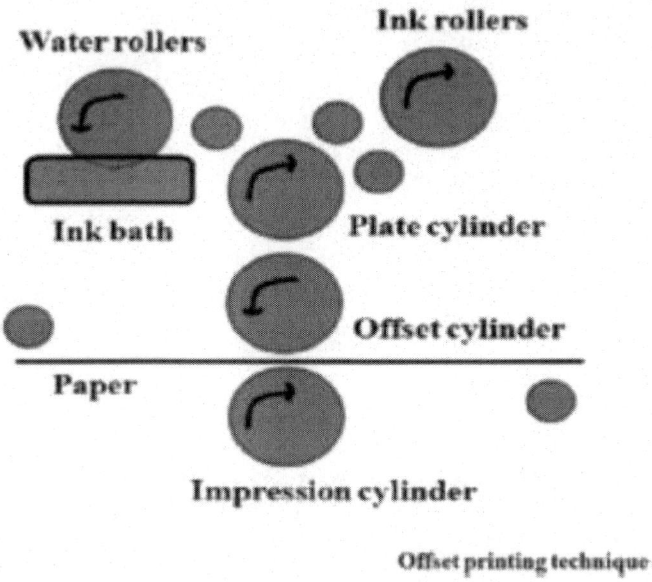

Figure 107. A Schematic drawing of offset printing technique [273].

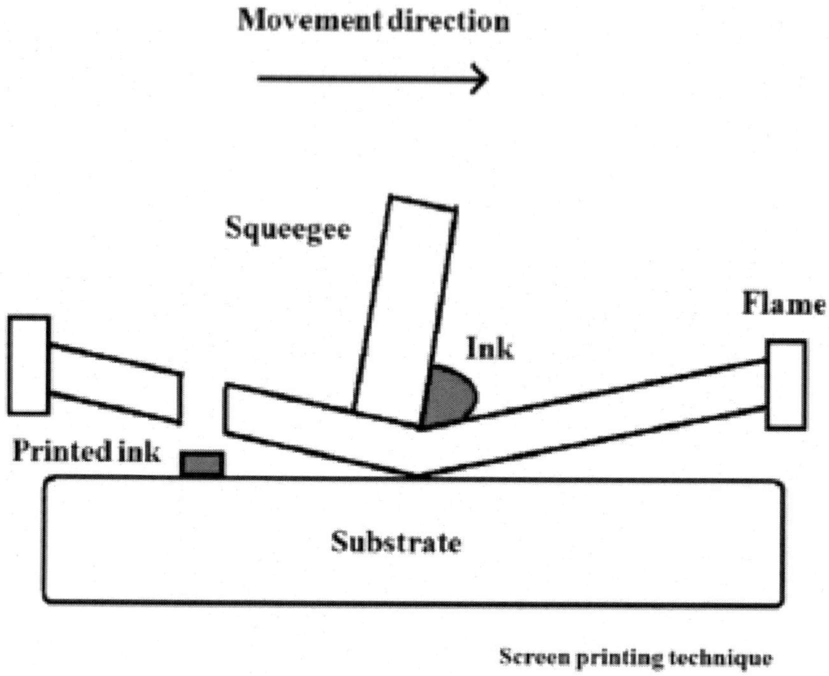

Figure 108. A schematic drawing of screen printing technique [273].

Screen printing was a very versatile printing technique that allowed for full 2-dimensional patterning of the printed layer. It was commonly used as a flat printing technique in batch processes, but it can also be adapted to a roll to roll (R2R) process, rotary printing, which enabled a rather high throughput. Generally, his technique used woven mesh to support an ink-blocking stencil. The attached stencil formed open areas of mesh that transferred ink or other printable materials, which can be pressed through the mesh as a sharp-edged image onto a substrate, as exhibited in Figure 108. A roller was moved across the screen stencil, pushing ink past the threads of the woven mesh in the open areas. Typically, ink solution required high viscosity of 1-50 PaS, the print resolution was around 100 μm; a wet layer thickness of up to 100 μm can be printed. The advantage of this technique was essentially no loss of coating solution during printing.

Inkjet printing was a relatively novel process from an industrial point of view for roll to roll (R2R) printing technology. The method has been driven forward by the typical low-cost inkjet printer for the home office. In this process, ink solution is generally kept in the cartridge connected to the printed head that is commonly fabricated by ceramic. It was well-known that this technique was versatile for several types of ink formulation with many different solvents. The computer controls the manipulation of this technique. It can also create a digital image by propelling droplets of ink onto substrate. The advantages of this technique were low cost, easily changeable digital print patterns and low material consumption. However, although inkjet printing for industry was R2R compatible, it offered a limited printing speed and unless the ink properties were properly tuned to a low viscosity (15 mPa S) and low evaporation rate, this technique commonly has problems with clogging, especially when small nozzles were used.

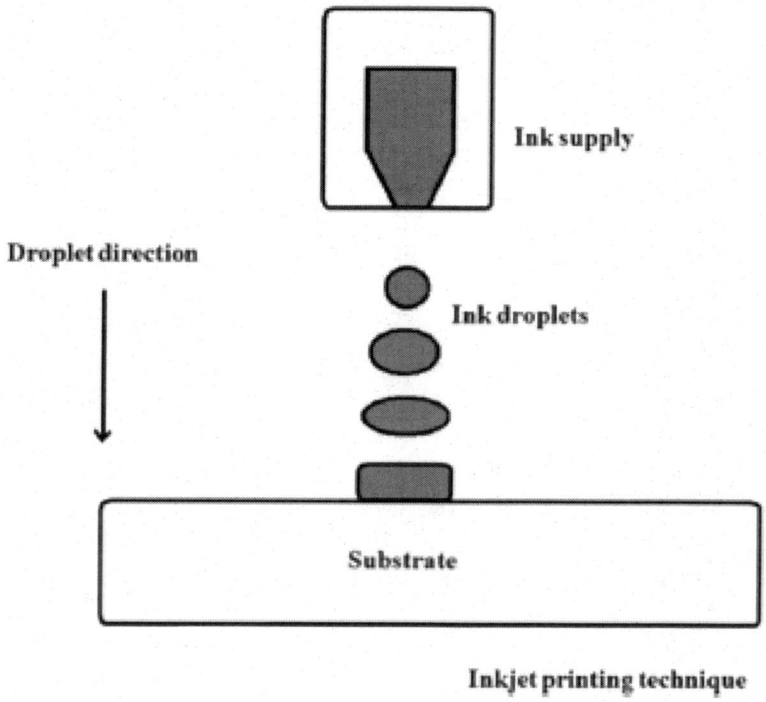

Figure 109. A schematic drawing of inkjet printing technique [273].

Up to the present time, many materials have been extensively investigated for feasibility of use with printed electronics. Conductive materials, which were generally required for all types of electronic components and conductivity, were considered depending on applications. Metals were considered if high and stable conductivity was required, while conductive polymers or organic molecules can sometimes be required if low conductivity was preferable. Generally, from an industrial point of view, the preparation of conductive materials can be categorized depending on type of materials such as metals, conductive polymers or organic molecules, etc.

The most common use of conductive materials involves metals. It is well-known that electronic applications, which required high and stable conductivity of 105 $S.cm^{-1}$, had to focus on these materials. Electronic components, commonly fabricated as thin film form, had to consider optimization of film thickness, conductivity and electronic efficiency. However, generally, what type of metals could be used as conductive materials in electronic components depended on numerous factors including requirement on conductivity, work function, oxidation stability in air as well as fabrication methods. Although metals offered various advantages including high and stable conductivity, the drawbacks on the high cost of materials and difficulty in processing had to be considered for scaling up to commercialization. The design metals for printed electronics had to consider the relationship of conductivity and nano-scale range of material shapes such as nano-tube, nano-particle, nano-wire as well as nano-rod, etc.

Metals have been used in electronics up to the present time. In particular, electrically conductive metals of both silver and copper nanoparticle ink patterns were fabricated as electronic components and they were consequently investigated for their electrical efficiency.

In 2012, Ummartyotin et al. successfully synthesized silver nanoparticles using conventional reaction of silver salt, using polyvinylidone (PVP) as stabilizer. This obtained silver nanoparticle led to further development of silver electrode for the organic light emitting diode system. Moreover, it was evident that the use of silver nanoparticle could be widely employed. It was consequently developed for optimization for environmental concerns as well as electrical efficiency. This was very interesting for the printed electronics community for electrically conductive polymers. However, although the use of silver nanoparticles offered numerous advantages in terms of electrical efficiency, the starting cost of this material was still high. From an industrial point of view, lower cost of preparation process should be applied for higher benefit. Therefore, numerous researchers have developed copper nanoparticles in order to solve this issue. Copper nanoparticles were prepared by both conventional reaction and high vacuum techniques. This preparation technique was similar to silver. However, electrical conductivity of copper was slightly decreased compared to silver, but the price of copper preparation made it more attractive for industrial purposes. Here, we wish to present the overview of the use metal nanoparticles for electrically conductive purposes. It provides several advantages for electronic devices, which required high electrical conductivity. However, printed electronics of conductive metals, crack formation, pinhole phenomena, adhesion, as well as compatibility with substrate, had to be well optimized if this conductive film was to be used with plastic substrate. These concerns may decrease the overall device performance.

Conductive polymers were theoretically defined as organic polymers that conducted electricity. Potential applications were numerous in devices such as fuel cells, supercapacitors, sensors, electrochromic displays as well as solar cells. Because the use of metals was considered environmentally harmful, conductive polymers were alternative candidates for replacing metal use. Moreover, the use of polymers has received more attention in terms of easier processability and cost effectiveness. This made conductive polymers interesting. In the past, electrically conductive polymers were first reported in 1977. The discovery of π- conjugated polymer polyacetylene could be made electrically, conducting by chemical doping. This led to further developments and the progress of this research field made conductive polymers including polyacetylene, poly (p-phenylene vinylene), polythiophene, polyaniline, polyphenylene sulfide, as well as polypyrrole, widely employed in many electronic industries. Conductive polymers have been developed to support research and academic communities. Morrin et al. investigated polyaniline film prepared by inkjet printers for electronic applications. Conductive film offered excellent properties of both physical characteristics as well as conductive efficiency. Moreover, Cho et al. proposed that this printing technique could enhance efficiency by inserting a vapor deposition polymerization step. This sub-process did not require additive as stabilizer and surfactant to form a conductive polymer pattern. This was still similar to developments for other conductive polymers such as poly(3,4-ethylenedioxy)-thiophene-poly(styrene sulfonate) or PEDOT: PSS, poly(phenylene-ethynylene) and poly(phenylene-vinylene).

Apart from metals and conductive polymers, other organic molecules such as carbon nanotube, graphite and graphene should be considered for their electrical conductivity in printed electronics. Carbon nanotube, theoretically defined as allotropes of carbon with a cylindrical nanostructure, have been extensively studied in order to explore their excellent electrical, thermal and mechanical properties. Although their electrical conductivity and small size were very interesting in terms of energy conversion, limited properties such as transparency and flexibility for electronic components had to be considered. The optical light transmittance was very low. Moreover, there may be concerns about the use carbon nanotube preparation costs and health safety reasons. Graphite and graphene are also excellent materials for printed electronics. It is commonly composed of carbon atoms linked together like a honeycomb network. These materials not only offered excellence with electrical conductivity but also with mechanical strength. This excellent correlation of both electrical and mechanical properties led to numerous applications, especially sensors, energy storage devices, and transparent conductive electrodes for solar cells and displays.

Semiconductors were commonly required for all fields of electronic components, especially field effect transistors. Its electrical conductivity intermediated in order of magnitude between conductor and insulator. This meant that conductivity was roughly exhibited in the range of 10^{-3} to 10^{-8} S/cm. In semiconductors, electric current was often schematized as being carried either by the flow of electrons to rich-part of positive charge or by the flow of positively charged or hole to rich-part of negative charge. This was totally different from the metallic conductor, which had a current carried only by the flow of electrons. Generally, semiconductors can be prepared by using two different materials. Inorganic semiconductors were excellent in terms of device performance and stability, whereas organic semiconductors typically had superior processability and flexibility. However, a hybrid system of inorganic and organic semiconductors, prepared by inorganic nanoparticles embedded in organic matrix, is sometimes required in order to optimize overall device performance.

Organic semiconductors are organic materials with semiconductor properties. Single molecules, oligomers as well as organic polymers can be made into semiconductors. Generally, in organic semiconductors, current carriers and mobility were holes and electrons in π-bonds. Electrons can move via π-electron cloud overlaps, especially by hopping and tunneling. When we fabricated as electronic component, it depended strongly on numerous factors including impurities, atmosphere, solvents, concentration, deposition technique, annealing conditions, geometry, roughness, film thickness, and interface with substrate, as well as applied electric field. In other words, this type of semiconductor can function perfectly if working function among electric donor molecules, acceptor molecules and relevant properties is appropriate. From an electrically printable point of view, organic semiconductors were preferably required. Significant performance of this material over inorganic semiconductors was made to be solution processable, which led to a more flexible process. However, the use of organic semiconductors had to be carefully considered for applications that required high air stability. This may lead to decreased device efficiency. Examples of widely use of organic semiconductors are poly(3,3'-didodecyl quaterthiophene) or PQT and poly(2,5-bis(3-tetradecyllthiophen-2-yl)thieno[3,2-b] thiophene) or PBTTT.

Although inorganic semiconductors often had better performance and stability than organic semiconductors, the processing technique and cost had to be considered for mass production process. The examples of inorganic semiconductors focused on the use of silicon or oxide of transition metal. Generally, only silicon film was fabricated by a chemical vapor deposition technique. This technique used high temperature (up to 400°C) for film nucleation, which obtains a good crystallinity and high electron mobility. Furthermore, this technique also allowed us to obtain film products with high purity and homogeneity. However, from the viewpoint of printed electronics, silicon film had to be used as nanoparticle form. The nano-scale size of silicon can be stored as ink solution in order to proceed in the printed electronics industry. It is important to note that this preparation technique offered less product quality, which resulted in less overall device performance.

Another class of inorganic semiconductors focused on oxides of transition metals. It was typically fabricated by radio frequency (RF) magnetron sputtering or pulsed laser deposition. This type of material was mainly prepared for transparent feature purpose. Examples of this material are ZnO and Ga_2O_3-In_2O_3-ZnO systems, and it was consequently fabricated in nanotube, nanorod and nanoparticle form. For the printed electronics industry, the ink solution of this transition metal oxide can be prepared by dissolving the oxide of the transition metal in solvent. Sometimes, hybrid composites between inorganic particles and the polymer matrix were prepared in order to optimize electrical performance and physical characteristics of materials.

Electrical insulating components are important in electronic devices in order to prevent shortage of crossing conductive wires. A high capacitance value was often desirable, not only in capacitors, but also in transistors to obtain low-voltage operation. This can be achieved by using a thin dielectric layer with a large permittivity. Also, it was important for the electronic device to prevent leakage currents or electrical shortages through the dielectric media, which commonly referred to dielectric layers that were thin, inhomogeneous and had a low dielectric strength.

In transistors, the dielectric properties at interface were great importance. This was probably because most of the charge transport in the semiconductor took place within a few nanometers from this interface. Whereas a high capacitance can be reached by using high permittivity gate dielectrics, the charge carrier mobility was reduced due to randomly oriented dipole moments in the gate dielectric causing a broadening of the density of states in the semiconductor near the interface. A classic group of materials can be used as dielectric in electronic devices. Inorganic materials such as silica and alumina were preliminary investigated due to thermal stability and relatively strong dielectric permittivity. However, the processing of inorganic semiconductors generally required high temperature (up to 400ºC) and this led to reduced efficiency if this material was expected to be scaled up for electronic industry. In order to solve this issue, numerous researchers have investigated insulating polymers such as PVP, PMMA, PET, PI, PP, PVA, PS for replacing inorganic semiconductor. Generally, polymers are typically good candidates. Most polymers are also inexpensive and available in large quantities.

Up to the present time, a compromise between the use of organic and inorganic semiconductors has received much attention in the electronic device industry. The composite between the two different materials was successfully prepared. The nanoparticle of the inorganic semiconductor was embedded into an organic matrix. This could solve the problematic concern on flexibility in printing technology.

Electronic components, generally fabricated as thin film form, were fundamentally necessary in order to fabricate electronic devices. From the theoretical point of view, electronic components can be categorized into two different classes. Single components such as conductive wires, resistors, inductors and antennas are considered passive components, whereas transistors, diodes, memories, batteries and solar cells are active components. Here, in order to understand printing technology for electronic devices, we wish to give a tentative overview on electronic components.

Over the last decade, transistors have moved from being a convenient tool for probing the charge transport in high band-gap, low conductivity organic semiconductors into an emerging technology enabling electronic functionalities to be integrated on electronic substrate for a range of applications. In the past, transistors were previously the main key component in most electronic systems 78. They were widely used, for example, as matrix controlling elements in displays and as logic gates in digital electronic circuits. A logic inverter can be made by simply connecting a transistor with a resistor. If the inverter has a gain (dV out/dV in > 1), it can be used as an amplifier or as a building block for more complicated circuits, such as rings, oscillators, flip-flops, ship registers, row decoders, and code generators, as well as transponder circuits.

Transistors have been continually developed up to the present time. Both inorganic and organic materials have been modified in order to tailor excellent properties of transistor. Commonly, in industry, the transistor has been used to fabricate liquid crystal displays (LCD) and light emitting diode (LED). Impressive progress has also been made in realizing logic circuits based on transistors for applications in radiofrequency identification tagging (RFID) and chemical or biological sensing. This rapid move towards real-world applications has been enabled by broad, interdisciplinary research efforts to better understand the materials, processing and device architecture requirements for semiconductors to achieve a higher level of transistor performance. Figure 110 exhibits the schematic diagram of transistor.

On the other hand, however, transistor performance not only depends on transistor type, materials and dimensions, but also on the deposition technique and substrate properties. The mobility of the semiconductor in transistor has been shown to strongly decrease with increasing roughness of the dielectric interface. This was a concern since print electronic commonly offered film with high roughness. While smooth surface and interface was preferred, this situation was actually far from reality. Kaihovirta et al. proposed an effective way to solve this issue by using a more robust transistor type, for example, by incorporating mobile ions in the device structure. It should be less sensitive to the surface roughness due to the homogeneous potential close to the dielectric/semiconductor interface, which is formed by the electric double layer. For marketing purposes on commercially available product, in order to take advantages of low cost in processing step, print electronics should be conducted. Numerous types of print transistors have been investigated. Both organic and inorganic transistors have been discussed.

Organic transistors have developed an emerging technology enabling electronic functionalities to be integrated on flexible substrate. Recently, printed electronics have been extensively developed in order to scale up to commercially available products. Significant changes on organic transistors for printed electronics have consequently improved to obtain excellent performance. This concept can lead to developments of novel organic materials for organic based electronic components. Up to the present time, numerous papers on organic transistors have been reviewed. Gelinck et al. reported the use of organic transistor as flexible displays and microelectronic application. Sirringhaus et al. reported device physics and stability. Furthermore, low voltage operation of organic transistor has been investigated. This was probably due to the fact that organic transistors were preferably used.

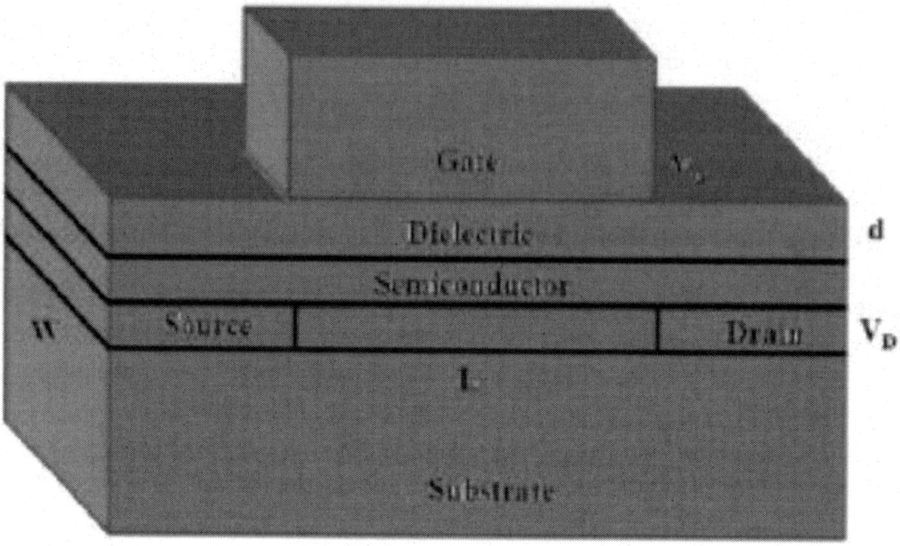

Figure 110. Schematic diagram of transistor [273].

After the first demonstration by Brody and Page research group, inorganic transistors have been initially fabricated. However, so far the devices generally fabricated by this inorganic material were inappropriate for mass production and they commonly offered a low quality of transistor characteristics that subsequently reduced transistor performance. Up to the present time, the limited flexibility of this inorganic transistor was still an issue.

Recently, Kim et al. have developed an inorganic transistor with the additional feature of flexibility. The concept of inorganic transistor development was continuously conducted. This is probably because inorganic material provides high stability in environments, superior to organic materials. The optimization between inorganic transistors commonly fabricated as thin film material for flexibility tailored excellent performance of the transistor. In this article, the connection of ultrathin Si-based logic inverters with flexible metallic serpentine structures was reported. This inverter also performed well after the bending and folding test. However, from the viewpoint of the printed electronics industry, this deposition technique for inorganic transistors was not feasible. Harting et al. reported that inorganic materials were commonly prepared as nanoparticle form. The 5-50 μm film thickness was prepared and printability was still being investigated. However, transistor performance was very poor, showing no saturation and a low on/off ratio. In order to achieve excellent transistors fabricated by inorganic materials with additional features of flexibility and stability, materials and relevant fabrication techniques needed to be developed.

In electronics, the diode was a type of two-terminal electronic component with non-linear resistance and conductance, distinguishing it from components such as two-terminal linear resistors, which obeyed Ohm's law. The most common use of diode is currently focused on semiconductor diodes, which are composed of a crystalline piece of semiconductor material connected to two electrical terminals. Up to the present time, numerous applications of diode have been developed, such as photovoltaic and organic light emitting diodes.

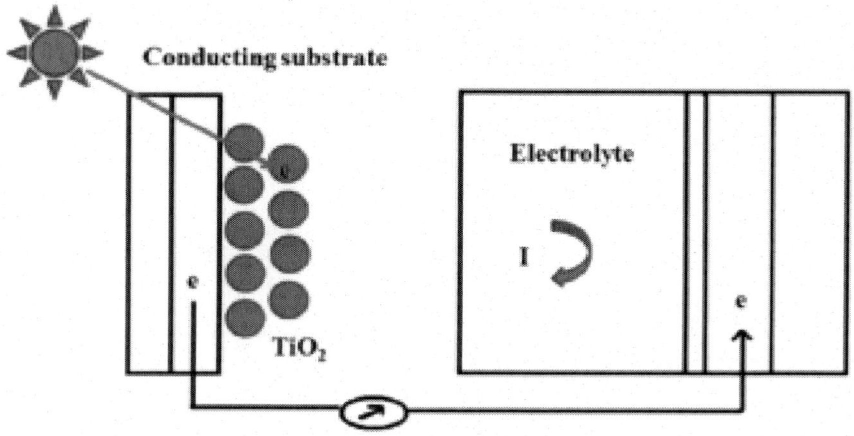

Figure 111. exhibits the schematic diagram of solar cell [273].

Photovoltaic was a promising future source of sustainable electricity production. It was developed as an alternative energy source. However, the main drawback of this technology concerned inorganic material, and the effective cost was still high. In order to scale up for commercialization, a significant cost reduction was required. Based on common use of photovoltaic material, dye solar cell was among the emerging solar technologies that have potential for low cost along with relatively high efficiency. However, from an industrial point of view, the roll-to-roll mass production of print electronic concepts could be employed. The photoelectrode consisted of a nanocrystalline TiO_2 layer deposited on substrate with a conducting layer, conventionally a transparent conducting oxide plastic or metal foil. The TiO_2 layer was sensitized with dye forming a monolayer that absorbed the light. The excitation of a dye molecule, by absorbing a photon from sunlight, which resulted in an electron injection into the conduction band of the semiconductor (TiO_2), the external circuit depicted this electron and transferred it toward the counter electrode. Figure 111 exhibits the schematic diagram of solar cell.

Organic light emitting diodes (OLEDs) have attracted considerable interest owing to their promising applications in flat-panel displays by replacing cathode ray tubes (CRTs) or liquid crystal displays (LCDs). Organic electroluminescent (OEL) materials, based on π-conjugated molecules, were almost insulators, and light was produced by recombination of holes and electrons, which had to be injected at the electrodes. The anode was transparent and usually prepared from indium tin oxide (ITO), while the cathode was reflective and prepared from metal. The thickness of the organic layer was very thin, between 100 and 150 nm. When voltage was applied between the electrodes, charges were injected in the organic material, holes from the anode and electrons from the cathode. Then, the charges move inside the material, generally by hopping processes, and then recombine to form excitons. The location of the recombination zone in the diode was a function of the charge mobility of the organic material as well as of the electric field distribution. After diffusion, the exciton recombined and a photon was consequently emitted. To the best of our knowledge, numerous efforts have been continuously developed for light emitting device. Our research group has successfully prepared organic light emitting diode by using cellulose based composite as flexible substrate. Figure 112 exhibits the schematic diagram of organic light emitting diode.

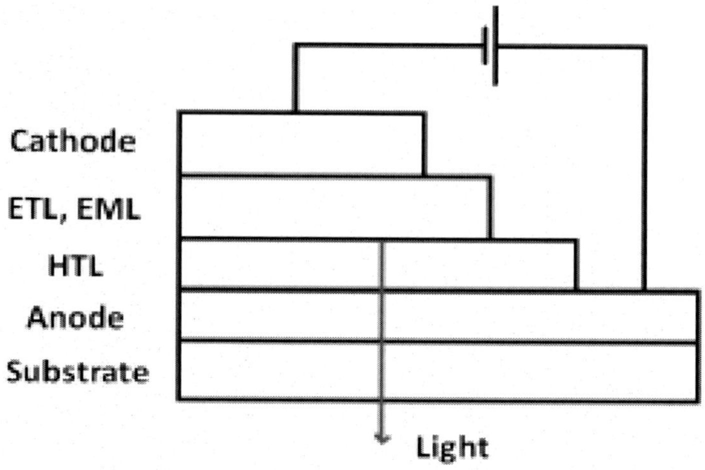

Figure 112. Schematic diagram of organic light emitting diode [273].

However, from the viewpoint of print electronics, ink solution was prepared as an active nanoparticle solution of ceramic or polymer. The combination of ceramic and polymer, called hybrid composite, is sometimes required. Although print electronics was an excellent technique for the roll-to-roll process, the optoelectronic quality of OLED was still under-developed. The push towards the print electronic method with high OLED quality needed to be further investigated.

Energy storage devices, generally referred to as batteries and supercapacitors, were made up of two electrodes separated by an electrolyte. The battery can either be non-rechargeable (primary) or rechargeable (secondary) form. The electric charge was stored as chemical energy through the electrochemical redox reaction in the electrode materials. However, for the supercapacitor, the electrostatic charge was stored through the formation of an electric double layer at the interface between the electrolyte and the electrodes. The battery typically had a larger energy density, while the charging and discharging speeds and the power density were lower. In supercapacitors, on the other hand, charging involved no phase change. Consequently, the charging and discharging speeds, as well as the cyclability, were high. Figure 113 exhibits the schematic diagram of the battery and supercapacitor.

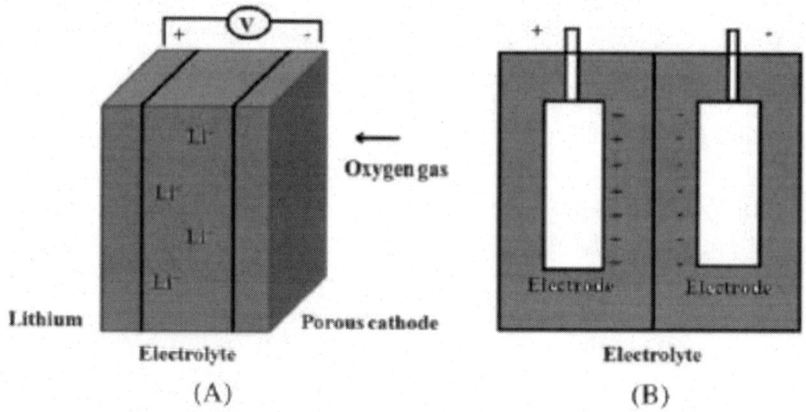

Figure 113. Schematic diagram of battery and supercapacitor [273].

For printable electronics, high surface roughness of print film can be easily obtained. While this could be detrimental for transistors and diode, it could be beneficial for energy storage devices, where a large surface area and electrolyte absorption was preferred. Up to the present time, mostly publications reported on the preparation of novel materials and their preparation for energy storage devices. However, these fabrication techniques of fabrication were high cost so it was difficult to scale up for mass production. The push for preparation of energy storage devices using the print electronic method must be investigated.

Electronic memory devices were theoretically required for storing digital information and used in applications such as radiofrequency identification (RFID) tags. Numerous papers have reported on the development of electronic memory devices. A non-volatile type of electronic memory device did not require power to maintain the stored information, whereas a volatile memory device lost the information when the power was not applied. Currently memory devices can be categorized into other subcategories, such as write-once read-many times (WORM) memories. However, in order to use electronic memory devices with high efficiency, writing/reading speed, retention time, on/off ratio, as well as yield of the fabrication process were important parameters that should be developed.

Base on the RFID system, the fabrication of material should focus on the correlation of electronic and magnetic properties. A wireless non-contact system that used radiofrequency electromagnetic fields to transfer data from a tag attached to an object was used for the purposes of automatic identification and tracking. The tags were powered by the electromagnetic fields used to read them. On the other hand, the local power source emitted radio waves. The tags contained electronically stored information which can be read from up to several meters away. Recently, the RFID system was extensively employed in many industries ranging from pharmaceutical science, livestock, as well as identity cards.

From the viewpoint of mass production industry, printable electronic has been widely employed for RFID system. However, this method is still in research, and development on RFID quality is still progressing in order to obtain high RFID performance. Figure 98 exhibits the schematic diagram of RFID.

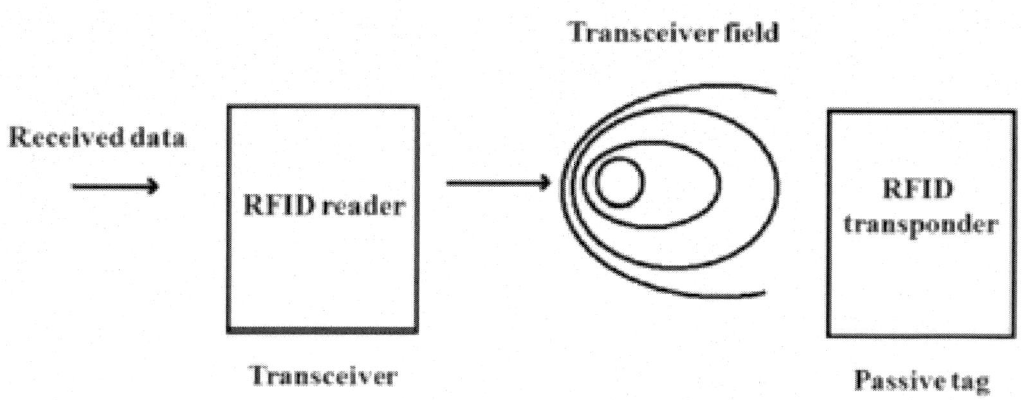

Figure 114. Schematic diagram of RFID system [273].

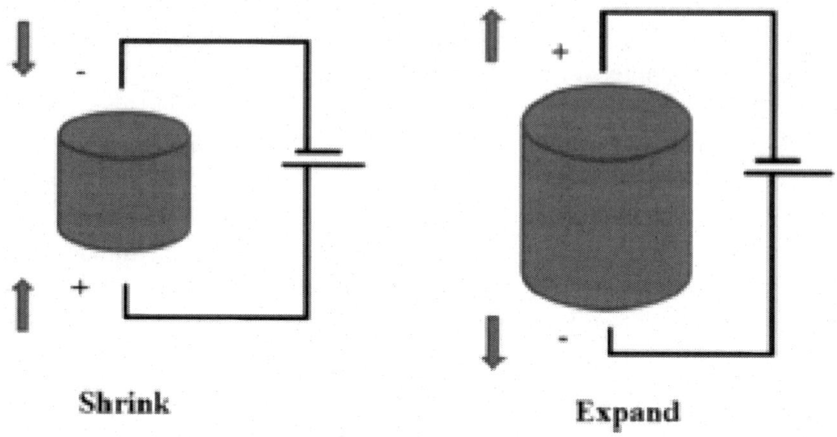

Figure 115. Schematic diagram of piezoelectric material [273].

Actuators have been extensively developed in numerous applications, ranging from micro-insect robots, microflying objects and micro-electromechanical systems. The theory of actuator was based on piezoelectric, which involved the correlation between electronic and mechanic. The charge that accumulated in certain solid materials in response to applied mechanical stress was investigated. In other words, this effect was understood as the linear electromechanical interaction between the mechanical and the electronic state in crystalline materials with no inversion symmetry. Moreover, this effect was considered a reversible process in that materials exhibited the direct piezoelectric effect (the internal generation of electrical charge resulting from an applied mechanical force) also exhibited the reverse piezoelectric effect (the internal generation of a mechanical strain resulting from an applied electrical field). Figure 115 exhibits the schematic diagram of piezoelectric material.

Chapter 6

REMAINING CHALLENGE AND FUTURE OUTLOOK

6.1. REMAINING CHALLENGE AND FUTURE OUTLOOK

As mentioned above, the use of cellulose presented excellence in terms of engineering properties; many efforts resulted in its development to replace many engineering materials. It was important to note that cellulose was interesting because if environmental preservation concerns. It is non-toxic and does not release hazardous compounds during processing and application. Moreover, the properties of cellulose were evident such as high stiffness, high aspect ratio, low coefficient of thermal expansion, as well as high chemical resistance. Cellulose therefore gained more and more interest in developments in other applications. Moreover, one of the challenges on the development on cellulose was cost effectiveness. Sources of cellulose were easy to discover. It was considered the most abundant naturally occurring biopolymer. In this scenario, the use of cellulose was not only interesting because of cost effectiveness and engineering properties, but because the concerns over hazardous substances and toxic materials could be overcome. This is probably why cellulose was being pushed in efforts to be developed as an engineering material. From the view point of flexible electronic, the role of cellulose was versatile due to interesting properties on high aspect ratio. Moreover, it will be interesting if print electronic is fabricated and therefore gains effective mass production in industrial commercialization. However, up to the present time, due to the nature of cellulose, it is important to note that the processing of cellulose requires a high amount of water, similar to the paper making process. To prevent the effect of high water absorption, it was a concern whether cellulose could be employed as electronic component. On this topic, in order to use cellulose in flexible electronic for lithium based battery, it should be noted that many aspects should be carefully considered.

6.1.1. Long-Term Stability

One of remaining challenges on the use of flexible electronic for lithium-based battery was its long-term stability. Although conductive cellulose was successfully designed and it satisfied the criteria on conductivity, the long-term stability was still an important concern. In industrial commercialization, long-term stability and reliability of materials should be preliminarily considered for mass production process. During operations, heat can be generated and moisture sensitivity of cellulose can shorten the lifetime of the electronic device. The design of conductive cellulose holds much interest due to the additional feature of flexibility and conductivity. However, the use of cellulose was still limited due to high water absorption ability, which may lead to shortening the service lifetime of the electronic device. Flexible electronic for lithium-based battery should focus on the design for long service lifetime of material and low cost of product. However, ideally, the concept on service lifetime of material should be considered to extend the performance of conductive cellulose for flexible electronic.

6.1.2. Reproducibility

Reproducibility was considered one important aspect for the mass production process. Before application in industrial commercialization, the result from the lab-scale level should be confirmed before consideration of mass production process. The use of bio-based materials should be considered. Cellulose was used as raw materials derived from plant-based cellulosic materials. To use cellulose with higher efficiency, the purification of cellulose has been extensively considered. Lignin and hemicellulose must be carefully controlled during purification process. Moreover, the amount of cellulose that can be extracted from a plant was varied due to plant species, soil condition, age of wood, as well as any environmental concern. Therefore, the raw materials for cellulose production should be consistent before utilization in industry. Furthermore, it is also easier to control if cellulose is further modified for conductivity by incorporating with conductive polymer and doped with metallic particles and active carbon. In industry, the design of conductive cellulose should require consistent quality control of product and process. Controlling properties of conductive cellulose can tentatively predict the service lifetime of device.

6.1.3. Characterization the Dispersion State

Characterization on the dispersion state of conductive cellulose was one of the important aspects to consider. Improving the conductivity of cellulose involves the incorporation of cellulose and conductive polymer and the insertion of small amounts of metallic particles and active carbon into cellulose. Characterization on dispersion state of conductive polymer, metallic particles as well as active carbon was therefore important in order to understand the morphological properties of materials and it can therefore be used to predict mechanical properties. In this aspect, characterization of morphological properties should be carefully considered. Application involved conductivity should involve concerns on the array of conductive particles while application involved dielectric and piezoelectric properties of materials should focus on the prevention of the conductive pathway of conductive particles.

REFERENCES

[1] Greffroy, B., Le Roy, P. & Prat, C. (2006). Review: Organic light emitting diode (OLED) technology: materials, devices and display technologies. *Polymer International*, *55*, p. 572-582.

[2] Thejokalyani, N. & Dhoble, S. J. (2014). Novel approaches for energy efficient solid-state lighting by RGB organic light emitting diodes- A review. *Renewable and Sustainable Energy Reviews*, *32*, p. 448-467.

[3] Thejokalyani, N. & Dhoble, S. J. (2012). Organic light emitting diodes: Energy saving lighting technology- A review. *Renewable and Sustainable Energy Reviews*, 16, p. 2696-2723.

[4] Chen, D., et al., (2015). Advances in transparent glass-ceramic phosphors for white light-emitting diodes-A review. *Journal of the European Ceramic Society*, *35*, p. 859-869.

[5] Bui, D. A. & Hauser, P. C. (2015). Analytical devices based on light-emitting diodes- a review of the state-of-the-art. *Analytica Chimica Acta*, *853*, p. 46-58.

[6] Chang, M. H., et al., (2012). Light emitting diodes reliability review. *Microelectronics Reliability*, *52*, p. 762-782.

[7] Nnorom, I. C., Osibanjo, O. & Ogwuegbu, M. O. C. (2011). Global disposal strategies for waste cathode ray tubes. *Resources, Conservation and Recycling*, *55*, p. 275-290.

[8] Rosink, J. J. W. M., et al., (2004). The hopping electron cathode for cathode ray tubes. *Advances in Imaging and Electron Physics*, *131*, p. 147-237.

[9] Oyama, T., et al., (1999). A new layer system of anti-reflective coating for cathode ray tubes. *Thin Solid Films*, *351*, p. 235-240.

[10] Yamazaki, E. (1998). Design and performance of shadow-mask color cathode ray tubes. *Advances in Imaging and Electron Physics*, *105*, p. 141-265.

[11] Hahm, S. G., et al., (2013). Liquid crystal alignment in advanced flat-panel liquid crystal displays. *Current Opinion in Chemical Engineering*, *2*, p. 71-78.

[12] Ma, J., Ye, X. & Jin, B. Structure and application of polarizer film for thin-film-transistor liquid crystal displays. *Displays*, *32*. 32, p. 49-57.

[13] Rashad, A. M. (2015). Recycled cathode ray tube and liquid crystal display glass as fine aggregate replacement in cementitious materials. *Construction and Building Materials*, In Press.

[14] Tai, Y. T., Pearn, W. L. & Chen, C. C. (2015). Product mix determination and capacity allocation for heterogeneous products in thin film transistor liquid crystal display manufacturing. *Computers & Industrial Engineering*, *80*, p. 201-211.

[15] Kim, J. M., et al., (2014). A liquid crystal display with consistent moving image quality regardless of viewing angles. *Optics & Laser Technology*, *60*, p. 34-40.

[16] Ma, J. & Xuan, L. (2013). Towards nanoscale molecular switch-based liquid crystal displays. *Displays*, *34*, p. 293-300.

[17] Buckley, A. (2013). Organic light-emitting diodes (OLEDs) Materials, devices and applications.

[18] Laaperi, A. (2013). Active matrix, organic light-emitting diodes (AMOLEDs) for displays. *Organic light-emitting diodes (OLEDs)*, p. 445-458.

[19] Shinar, J. & Shinar, R. (2011). An overview of organic light-emitting diodes. *Comprehensive Nanoscience and Technology*, *1*, p. 73-107.

[20] Amendola, V., et al., (2006). Light-emitting molecular devices based on transition metals. *Coordination Chemistry Reviews*, *250*, p. 273-299.

[21] Shirota, Y. & Kageyama, H. (2013). Small molecular weight materials for (opto)electronic applications: overview, ed. H.o.O.M.f.O.a.O.e. Devices.

[22] Bousquet, A., et al., (2014). Conjugated-polymer grafting on inorganic and organic substrates: A new trend in organic electronic materials. *Progress in Polymer Science*, *39*, p. 1847-1877.

[23] Gao, Y. (2010). Surface analytical studies of interfaces in organic semiconductor devices. *Materials Science and Engineering: R: Reports*, *68*, p. 39-87.

[24] Lee, C. W., Kim, O. Y. & Lee, J. Y. (2014). Organic materials for organic electronic devices. *Journal of Industrial and Engineering Chemistry*, 2014. *20*, p. 1198-1208.

[25] Lee, W., Seo, J. H. & Woo, H. Y. (2013). Conjugated polyelectrolytes: A new class of semiconducting material for organic electronic devices. *Polymer*, *54*, p. 5104-5121.

[26] Brosseua, C. (2011). Emerging technologies of plastic carbon nanoelectronics: A review. *Surface and Coatings Technology*, *206*, p. 753-758.

[27] Nagao, M. & Yoshida, T. (2015). Fabrication of gated nano electron source for vacuum nanoelectronics. *Microelectronic Engineering*, *132*, p. 14-20.

[28] Shadman, F. (2012). Environmental challenges in nanoelectronics manufacturing. *Current Opinion in Chemical Engineering*, *1*, p. 258-268.

[29] Kajd, S., et al., (2010). Applications of biological pores in nanomedicine, sensing, and nanoelectronics. *Current Opinion in Biotechnology*, *21*, p. 439-476.

[30] Chen, Y. (2015). Nanofabrication by electron beam lithography and its applications: A review. *Microelectronics Engineering*, *135*, p. 57-72.

[31] Helfrich, W. & Schneider, W. G. (1995). Recombination radiation in anthracene crystals. *Physical review letters*, *14*, p. 229.

[32] Tang, C. W. & van Slyke, S. A. (1987). Organic electroluminescent diodes. *Applied Physics Letters*, *51*, p. 913.

[33] Chen, D., et al., (2015). Advances in transparent glass-ceramic phosphors for white light-emitting diodes- A review. *Journal of the European Ceramic Society*, *35*, p. 859-869.

[34] Zhou, G., Wong, W. Y. & Suo, S. (2010). Recent progress and current challenges in phosphorescent white organic light-emitting diodes (WOLEDs). *Journal of Photochemistry and Photobiology C: Photochemistry Reviews*, *11*, p. 133-156.

[35] Wang, J., et al., (2013). Key issues and recent progress of high efficient organic light-emitting diodes. *Journal of Photochemistry and Photobiology C: Photochemistry Reviews*, *17*, p. 69-104.

[36] Choi, M. C., Kim, Y. & Ha, C. S. (2008). Polymers for flexible displays: From material selection to device applications. *Progress in Polymer Science*, *33*, p. 581-630.

[37] Wang, P. C. & MacDiamid, A. G. (2007). Integration of *polymer*-dispersed liquid crystal composite with conducting *polymer* thin films toward the fabrication of flexible display devices. *Displays*, *28*, p. 101-104.

[38] Sierros, K. A., et al., (2010). Pulsed laser deposition of indium tin oxide films on flexible polyethylene naphathalate display substrates at room temperature. *Thin Solid Films*, *518*, p. 2623-2627.

[39] D'&rade, B. W., Kattamis, A. Z. & Murphy, P. F. (2015). Flexible organic electronic devices on metal foil substrates for light, photovoltaic, and other applications, ed. H.o.F.O. Electronics.

[40] Luo, C., et al., (2015). Roll-to-roll fabrication of organic nanorod electrodes for sodium ion batteries. *Nano Energy*, *13*, p. 537-545.

[41] Roth, B., Sondergaard, R. R. & Krebs, F. C. (2015). Roll-to-roll printing and coating techniques for manufacturing large-area flexible organic electronics, ed. H.o.F.O. Electronics.

[42] Krebs, F. C. (2009). Fabrication and processing of polymer solar cells: A review of printing and coating techniques. *Solar Energy Materials and Solar Cells*, *93*, p. 394-412.

[43] Chang, P. L., Wu, C. C. & Leu, H. J. (2012). Investigation of technological trends in flexible display fabrication through patent analysis. *Displays*, *33*, p. 68-73.

[44] Wakeham, S. J., et al., (2009). Low temperature remote plasma sputtering of indium tin oxide for flexible display applications. *Thin Solid Films*, *518*, p. 1355-1358.

[45] Grego, S., et al., (2007). A method to evaluate mechanical performance of thin transparent films for flexible displays. *Thin Solid Films*, *515*, p. 4745-4752.

[46] Sanghvi, A. P. (1991). Power shortages in developing countries: Impacts and policy implications. *Energy Policy*, *19*, p. 425-440.

[47] Jamieson, G. A. & Vicente, K. J. (2001). Ecological interface design for petrochemical applications: supporting operator adaptation, continuous learning, and distributed, collaborative work. *Computers & Chemical Engineering*, *25*, p. 1055-1074.

[48] Alazemi, J. & Andrews, J. (2015). Automotive hydrogen fuelling stations: An international review. *Renewable and Sustainable Energy Reviews*, *48*, p. 483-499.

[49] Najafi, S. K. (2013). Use of recycled plastics in wood plastic composites- A review. Waste Management, *33*, p. 1898-1905.

[50] Kosny, J., et al., (2014). A review of high R-value wood framed and composite wood wall technologies using advanced insulation techniques. *Energy and Buildings*, *72*, p. 441-456.

[51] Hetzer, M. & Kee, D. D. (2008). Wood/polymer/nanoclay composites, environmentally friendly sustainable technology: A review. *Chemical Engineering Research and Design*, *86*, p. 1083-1093.

[52] Ardanuy, M., Claramunt, J. & Filho, R. D. T. (2015). Cellulosic fiber reinforced cement-based composites: A review of recent research. *Construction and Building Materials*, *79*, p. 115-128.

[53] Wu, C. (2010). Production and characterization of optically transparent nanocomposite film, in Faculty of Forestry. 2010, University of Toronto: Toronto.
[54] Ander, P. (1994). The cellobiose-oxidizing enzymes CBQ and CbO as related to lignin and cellulose deradation- a review. *FEMS Microbiology Reviews*, *13*, p. 197-311.
[55] Michailof, C., et al., (2014). Quantitative and qualitative analysis of hemicellulose, cellulose and lignin bio-oils by comprehensive two-dimensional gas chromatography with time-of-flight mass spectrometry. *Journal of Chromatography A*, *1369*, p. 147-160.
[56] Chen, W. H. & Kuo, P. C. (2011). Isothermal torrefaction kinetics of hemicellulose, cellulose, lignin and xylan using thermogravimetric analysis. *Energy*, *36*, p. 6451-6460.
[57] Legnani, C., et al., (2008). Bacterial cellulose membrane as flexible substrate for organic light emitting devices. *Thin Solid Films*, *517*, p. 1016-1020.
[58] Okahisa, Y., et al., (2009). Optically transparent wood-cellulose nanocomposite as a base substrate for flexible organic light-emitting diode *displays*. *Composites Science and Technology*, *69*, p. 1958-1961.
[59] Ummartyotin, S. (2012). et al., Development of transparent bacterial cellulose nanocomposite film as substrate for flexible organic light emitting diode (OLED) display. *Industrial Crops and Products*, *35*, p. 92-97.
[60] Shi, Z., et al., (2013). Utilization of bacterial cellulose in food. Food Hydrocolloids, Article In Press.
[61] Hu, W., et al., (2014). Functionalized bacterial cellulose derivatives and nanocomposites. *Carbohydrate Polymer*s, *101*, p. 1043-1060.
[62] Bae, S., Sugano, Y. & Shoda, M. (2004). Improvement of bacterial cellulose production by addition of agar in a jar fermentor. *Journal of Bioscience and Bioengineering*, *97*, p. 33-38.
[63] Klemm, D., et al., (2001). Bacterial synthesized cellulose-artificial blood vessels for microsurgery. *Progress in Polymer Science*, *26*, p. 1561-1603.
[64] Crawford, G. (2015). Flexible flat panel displays.
[65] Brinker, C. J., et al., (1992). Review of sol-gel thin film formation. *Journal of Non-Crystalline Solids*, 147-148, p. 424-436.
[66] Colgan, E. G. (1990). A review of thin-film aluminide formation. *Materials Science Reports*, *5*, p. 1-44.
[67] Adams, D. P. (2015). Reactive multilayers fabricated by vapor deposition: A critical review. *Thin Solid Films*, *576*, p. 98-128.
[68] Dorval Dion, C. A. & Tavares, J. R. (2013). Photo-initiated chemical vapor deposition as a scalable particle functionalization technology (a practical review). *Powder Technology*, *239*, p. 484-491.
[69] Danafar, F., et al., (2009). Fluidized bed catalytic chemical vapor deposition synthesis of carbon nanotubes-A review. *Chemical Engineering Journal*, *155*, p. 37-48.
[70] Helmersson, U., et al., (2006). Ionized physical vapor deposition (IPVD): A review of technology and applications. *Thin Solid Films*, *513*, p. 1-24.
[71] Fernandez-Seara, J. & Pardinas, A. A. (2014). Refrigerant falling film evaporation review: Description, fluid dynamics and heat transfer. *Applied Thermal Engineering*, *64*, p. 155-171.
[72] Yildirim Erbil, H. (2012). Evaporation of pure liquid sessile and spherical suspended srops: A review. *Advances in Colloid and Interface Science*, *170*, p. 67-86.

[73] Tallaire, A., et al., (2013). Growth of large size diamond single crystals by plasma assisted chemical vapour deposition: Recent achievements and remaining challenges. *Comptes Rendus Physique*, *14*, p. 169-184.

[74] Okada, K. (2007). Plasma-enhanced chemical vapor deposition of nanocrystalline diamond. *Science and Technology of Advanced Materials*, *8*, p. 624-634.

[75] Stevens, M. P. (1999). *Polymer* Chemistry: An Introduction.

[76] Rubinstein, M. & Colby, R. H. *Polymer* Physics.

[77] Deng, H., et al., (2013). Progress on the morphological control of conductive network in conductive polymer composites and the use as electroactive multifunctional materials. *Progress in Polymer Science*, 2013. In Press.

[78] Abu Seman, M. N., et al., (2010). Reduction of nanofitration membrane fouling by UV-initiated graft polymerization technique. *Journal of Membrane Science*, *355*, p. 133-141.

[79] Ebara, M., et al., (2007). Surface modification of microfluidic channels by UV-mediated graft polymerization of non-fouling and smart polymers. *Radiation Physics and Chemistry*, *76*, p. 1409-1413.

[80] Wong, I. L., Eames, P. C. & Perera, R. S. (2007). A review of transparent insulation systems and the evaluation of payback period for building applications. *Solar Energy*, *81*, p. 1058-1071.

[81] Chopra, K. L., Major, S. & Pandya, D. K. (1983). Transparent conductors-A status review. *Thin Solid Films*, *102*, p. 1-46.

[82] Guo, C. F. & Ren, Z. (2015). Flexible transparent conductors based on metal nanowire networks. *Materials Today*, *18*, p. 143-154.

[83] Wang, Z. B., Helander, M. G. & Lu, Z. H. (2013). Transparent conducting thin films for OLEDs.

[84] Hu, J., et al., (2009). Design and validation of a low cost surface plasmon resonance bioanalyzer using microprocessors and a touch-screen monitor. *Biosensors and Bioelectronics*, *24*, p. 1974-1978.

[85] Burgi, L., et al., (2006). Optical proximity and touch sensors based on monolithically integrated polymer photodiodes and polymer LEDs. *Organic Electronics*, *7*, p. 114-120.

[86] Cho, N. H., et al., (2015). Quantitative assessment of touch-screen panel by nondestructive inspection with three-dimensional real-time display optical coherence tomography. *Optics and Lasers in Engineering*, *68*, p. 50-57.

[87] Piper, A. M. & Hollan, J. D. (2013). Supporting medical communication for older patients with a shared touch-screen computer. *International Journal of Medical Informatics*, *82*, p. 242-250.

[88] Gould, J. D., et al., (1990). Using a touchscreen for simple tasks. *Interacting with Computers*, *2*, p. 59-74.

[89] Sears, A. & Shneiderman, B. (1991). High precision touchscreen: design strategies and comparisons with a mouse. *International Journal of Man-Machine Studies*, *34*, p. 593-613.

[90] Mathew, X., et al., (2003). Development of CdTe thin films on flexible substrates- a review. *Solar Energy Materials and Solar Cells*, *76*, p. 293-303.

[91] Acuautla, M., et al., (2014). Direct laser patterning of a gas sensor on flexible substrate. *Procedia Engineering*, *2014*, p. 899-902.

[92] Kim, J. H. & Park, J. W. (2013). Improving the flexibility of large-area transparent conductive oxide electrodes on polymer substrates for flexible organic light emitting diodes by introducing surface roughness. *Organic Electronics*, *14*, p. 3444-3452.

[93] Soz, C. K., Yilgor, E. & Yilgor, I. (2015). Influence of the average surface roughness on the formation of superhydrophobic polymer surfaces through spin-coating with hydrophobic fumed silica. *Polymer*, *62*, p. 118-128.

[94] Lee, W. L., et al., (2013). Manufacture of optical substrates with high gas resistance as well as antistatic capability and their applications for flexible organic light emitting diodes (flexible OLEDs). *Materials Chemistry and Physics*, *142*, p. 701-706.

[95] Ouyang, F. Y. & Tai, W. L. (2013). Enhanced corrosion resistance of TiN-coated stainless steels for the application in flexible dye-sensitized solar cells. *Applied Surface Science*, *276*, p. 563-570.

[96] Zhang, B., et al., (2015). Enhancing fatigue cracking resistance of nanocrystalline Cu films on a flexible substrate. *Materials Science and Engineering: A*, *627*, p. 61-64.

[97] Tamagaki, H., Ikari, Y. & Ohba, N. (2014). Roll-to-roll sputter deposition on flexible glass substrates. *Surface and Coating Technology*, *241*, p. 138-141.

[98] Frisk, L. & Saarinen-Pulli, K. (2014). Reliability of adhesive jointed thinned chips on flexible substrates under humid conditions. *Microelectronics Reliability*, *54*, p. 2058-2063.

[99] Molina-Lopez, F., et al., (2014). Study of bending reliability and electrical properties of platinum lines on flexible polyimide substrates. *Microelectronics Reliability*, *54*, p. 2542-2549.

[100] Fahlteich, J., et al., (2009). Permeation barrier properties of thin oxide films on flexible polymer substrates. *Thin Solid Films*, *517*, p. 3075-3080.

[101] Baschek, G., Hartwig, G. & Zahradnik, F. (1999). Effect of water absorption in *polymer*s at low and high temperatures. *Polymer*, *40*, p. 3433-3441.

[102] Zhu, C. G., et al., (2014). Ultra reliable infrared absorption water vapor detection through the all-electronic feedback stabilization. *Optics Communications*, *315*, p. 388-393.

[103] Gozalvez-Zafrilla, J. M., et al., (2011). Fluid dynamic modeling of oxygen permeation through mixed ionic-electronic conducting membranes. *Journal of Membrane Science*, *378*, p. 290-300.

[104] Kim, S., et al., (1998). Diffusion and surface exchange coefficients in mixed ionic electronic conducting oxides from the pressure dependence of oxygen permeation. *Solid State Ionics*, *106*, p. 189-195.

[105] Chen, Y., et al., (2014). Influence of sealing materials on the oxygen permeation fluxes of some typical oxygen ion conducting ceramic membranes. *Journal of Membrane Science*, *470*, p. 102-111.

[106] Bo, S. F., et al., (2014). Fabrication of tunable [Al2O3:Alucone] thin-film encapsulations for top-emitting organic light-emitting diodes with high performance optical and barrier properties. *Organic Electronics*, *15*, p. 2546-2552.

[107] Qiang, Y. Y., et al., (2014). High barrier properties of transparent thin-film encapsulations for top emission organic light-emitting diodes. *Organic Electronics*, *15*, p. 1120-1125.

[108] Park, J., et al., (2012). Optical and barrier properties of thin-film encapsulations for transparent OLEDs. *Organic Electronic*s, *13*, p. 1956-1961.

[109] Ham, H., Park, J. & Kim, Y. (2011). Thermal and barrier properties of liquid getter-filled encapsulations for OLEDs. *Organic Electronics*, *12*, p. 2174-2179.

[110] Ogino, A. & Nagatsu, M. (2007). Gas barrier properties of hydrogenated amorphous carbon films coated on polymers by surface-wave plasma chemical vapor deposition. *Thin Solid Films*, *515*, p. 3597-3601.

[111] Seo, S. W., et al., (2014). Bending properties of organic-inorganic multilayer moisture barriers. *Thin Solid Films*, *550*, p. 742-746.

[112] Cho, A. R., et al., (2014). Flexible OLED encapsulated with gas barrier film and adhesive gasket. *Synthetic Metals*, *193*, p. 77-80.

[113] Choi, M. C., Kim, Y. & Ha, C. S. (2008). Polymer for flexible displays: From material selection to device applications. *Progress in Polymer Science*, *33*, p. 581-630.

[114] Oostra, A. J., Blom, P. W. M. & Michels, J. J. (2014). Prevention of short circuits in solution-processed OLED devices. *Organic Electronics*, *15*, p. 1166-1172.

[115] Leterrier, Y., et al., (2010). Mechanical failure analysis of thin film transistor devices on steel and polyimide substrates for flexible display applications. *Engineering Fracture Mechanics*, *77*, p. 660-670.

[116] Fenter, P., et al., (1997). Thermally induced failure mechanisms of organic light emitting device structures probed by X-ray specular reflectivity. *Chemical Physics Letters*, *277*, p. 521-526.

[117] Brockbank, S. A., et al., (2013). Infinite dilution activity coefficients and Henry's law constants of compounds in water using the inert gas stripping method. *Fluid Phase Equilibria*, *3448*, p. 45-51.

[118] Chen, F., et al., (2012). Henry's law constants of chlorinated solvents at elevated temperatures. *Chemosphere*, *86*, p. 156-165.

[119] Sahoo, S., Patel, S. & Mishra, B. K. (2011). Quantitative structure property relationship for Henry's law constant of some alkane isomers. *Thermochimica Acta*, *512*, p. 273-277.

[120] Allou, L., El Maimouni, L. & Le Calve, S. (2011). Henry's law constant measurements for formaldehyde and benzaldehyde as a function of temperature and water composition. *Atmospheric Environment*, *45*, p. 2991-2998.

[121] Sindo, K. (1996). Transport phenomena and materials processing.

[122] Shaegh, S. A. M., Nguyen, N. T. & Chan, S. H. (2011). A review on membraneless laminar flow-based fuel cells. *International Journal of Hydrogen Energy*, *36*, p. 5675-5694.

[123] Bishop, A. A. (1980). Transient friction and form loss factors in turbulent and laminar flow; a brief review. *Nuclear Engineering and Design*, *62*, p. 361-363.

[124] Kutilek, M. (1972). Non-Darcian Flow of water in soils_ Laminar region: A review. *Developments in Soli Science*, *2*, p. 327-340.

[125] Goulet, M. A. & Kjeang, E. (2014). Co-laminar flow cells for electrochemical energy conversion. *Journal of Power Sources*, *260*, p. 186-196.

[126] Khosla, P. K. & Abdallah, S. (2006). A review of streamline coordinate approach for in viscid flows. *Computers & Fluids*, *35*, p. 670-675.

[127] Paddock, S. & Jones, P. (2010). Streamlining enteral feeding: An overview of the process of reviewing and changing enteral feeding practices. *Australian Critical Care*, *23*, p. 42.

[128] Harmand, S., et al., (2013). Review of fluid flow and convective heat transfer within rotating disk cavities with impinging jet. *International Journal of Thermal Sciences*, *67*, p. 1-30.

[129] Wu, C. Y., et al., (2012). Organosilicon/silicon oxide gas barrier structure encapsulated flexible plastic substrate by using plasma-enhanced chemical vapor deposition. *Surface and Coating Technology*, *206*, p. 4685-4691.

[130] Korner, L., Sonnenfeld, A. & Rudolf von Rohr, P. (2010). Silicon oxide diffusion barrier coatings on polypropylene. *Thin Solid Films*, *518*, p. 4840-4846.

[131] Jin, S. B., et al., (2010). Characterization of silicon oxide gas barrier films with controlling to the ion current density (ion flux) by plasma enhanced chemical vapor deposition. *Thin Solid Films*, *518*, p. 6385-6389.

[132] So, Y. H., et al., (2011). Formation and photoluminescence of Si nanocrystals in controlled multilayer structure comprising of Si-rich nitride and ultrathin silicon nitride barrier layers. *Thin Solid Films*, *519*, p. 5408-5412.

[133] Wan, Z., et al., (2011). A ultra-thin silicon nitride barrier layer implementation for silicon quantum dots in amorphous silicon carbide matrix in photovoltaic application. *Energy Procedia*, *10*, p. 271-281.

[134] Chen, T. N., et al., (2006). Effects of plasma pretreatment on silicon nitride barrier films on polycarbonate substrates. *Thin Solid Films*, *514*, p. 188-192.

[135] Xia, H., et al., (2015). Mechanical and thermal properties of reduced graphene oxide reinforced aluminum nitride ceramic composites. *Materials Science and Engineering: A*, *639*, p. 29-36.

[136] Zhang, F., et al., (2004). Preparation of nanocrystalline ceramic oxide powders in the presence of anionic starburst dendrimer. *Materials Letters*, *58*, p. 3285-3289.

[137] Girtan, M., et al., (2013). On the properties of aluminum doped zinc oxide thin films deposited on plastic substrates from ceramic targets. *Applied Surface Science*, *274*, p. 306-313.

[138] Lavoine, N., et al., (2012). Microfibrillated cellulose- Its barrier properties and applications in cellulosic materials: A review. *Carbohydrate Polymers*, *90*, p. 735-764.

[139] Kulpinski, P. (2005). Cellulose nanofibers prepared by the N-methylmorpholine-N-oxide method. *Journal of Applied Polymer Science*, *98*, p. 1473-1482.

[140] Ummartyotin, S. & Manuspiya, H. (2015). A critical review on cellulose: From fundamental to an approach on sensor technology. *Renewable and Sustainable Energy Reviews*, *41*, p. 402-412.

[141] Obama, P., et al., (2012). Combination of enzymatic hydrolysis and ethanol organosolv pretreatments: Effect on lignin structures, delignification yields and cellulose-to-glucose conversion. *Bioresource Technology*, *112*, p. 156-163.

[142] Fernandez-Bolanos, J., et al., (1999). Characterization of the lignin obtained by alkaline delignification and of the cellulose residue from steam-exploded olive stones. *Bioresource Technology*, *68*, p. 121-132.

[143] Liu, R., et al., (2014). Comparison on properties of lignocellulosic flour/polymer composites by using wood, cellulose, and lignin flours as fillers. *Composites Science and Technology*, *103*, p. 1-7.

[144] Thoorens, G., et al., (2014). Microcrystalline cellulose, a direct compression binder in a quality by design environment- A review. *International Journal of Pharmaceutics*, *473*, p. 64-72.

[145] Collard, F. X. & Blin, J. (2014). A review on pyrolysis of biomass constituents: Mechanisms and composition of the products obtained from the conversion of cellulose, hemicelluloses and lignin. *Renewable and Sustainable Energy Reviews*, *38*, p. 594-608.

[146] Lede, J. (2012). Cellulose pyrolysis kinetics: An historical review on the existence and role of intermediate active cellulose. *Journal of Analytical and Applied Pyrolysis*, *94*, p. 17-32.

[147] Abdul Khalil, H. P. S., B. A. H. & Ireana Yusra, A. F. (2012). Green composites from sustainable cellulose nanofibrils: A review. *Carbohydrate Polymers*, *87*, p. 963-979.

[148] Shah, N., et al., (2013). Overview of bacterial cellulose composites: A multipurpose advanced material. Carbohydrate *Polymers*, *98*, p. 1585-1598.

[149] Vo, L. T. T., et al., (2013). All-cellulose composites from woven fabrics. Composites Science and Technology, *78*, p. 30-40.

[150] HUber, T., et al., (2012). Solvent infusion processing of all-cellulose composite materials. *Carbohydrate Polymers*, *90*, p. 730-733.

[151] Abdul Khalil, H. P. S., et al., (2014). Production and modification of nanofibrillated cellulose using various mechanical processes: A review. *Carbohydrate Polymers*, *99*, p. 649-665.

[152] Meguid, M. A., et al., (2008). Physical modeling of tunnels in soft ground: A review. Tunnelling and Underground Space Technology, *23*, p. 185-198.

[153] Kopac, J. & Krajnik, P. (2006). High-performance grinding- A review. *Journal of Materials Processing Technology*, *175*, p. 278-284.

[154] Ahmed, F. E., Lalia, B. S. & Hashaikeh, R. (2015). A review on electrospinning for membrane fabrication: Challenges and applications. *Desilination*, *356*, p. 15-30.

[155] Braghirolli, D. I., Steffens, D. & Pranke, P. (2014). Electrospinning for regenerative medicine: a review of the main topics. *Drug Discovery Today*, *19*, p. 743-753.

[156] Chronakis, I. S. (2005). Novel nanocomposites & nanoceramics based on *polymer* nanofibers using electrospinning process- A review. *Journal of Materials Processing Technology*, *167*, p. 283-293.

[157] Gamelas, J. A. F., et al., (2015). On the morphology of cellulose nanofibrils obtained by TEMPO-mediated oxidation and mechanical treatment. *MIcron*, *72*, p. 28-33.

[158] Besbes, I., Alila, S. & Boufi, S. (2011). Nanofibrillated cellulose from TEMPO-oxidized eucalyptus fibres: Effect of the carboxyl content. Carbohydrate *Polymers*, *84*, p. 975-983.

[159] Lavazza, M., et al., (2011). Oxidation of galactomannan by laccase plus TEMPO yields an elastic gel. *Journal of Biotechnology*, *156*, p. 108-116.

[160] Saghir, S., et al., (2008). Structure characterization and carboxymethylation of arabinoxylan isolated from Ispaghula (Plantago ovata) seed husk. *Carbohydrate Polymers*, *74*, p. 309-317.

[161] Setyawati, M. I., Chien, L. J. & Lee, C. K. (2009). Self-immobilized recombinant Acetobacter xylinum for biotransformation. *Biochemical Engineering Journal*, *43*, p. 78-84.

[162] Embuscado, M. E., Marks, J. S. & BeMiller, J. N. (1994). Bacterial cellulose. I. Factors affecting the production of cellulose by Acetobacter xylinum. *Food Hydrocolloids*, *8*, p. 407-418.

[163] Embuscado, M. E., BeMiller, J. N. & Marks, J. S. (1996). Isolation & partial characterization of cellulose produced by Acetobacter xylinum. *Food Hydrocolloids*, *10*, p. 75-82.

[164] Ion, V. A., Parvulescu, O. C. & Dobre, T. (2015). Volatile organic compounds adsorption onto neat and hybrid bacterial cellulose. *Applied Surface Science*, *335*, p. 137-146.

[165] Mohammadkazemi, F., Azin, M. & Ashori, A. (2015). Production of bacterial cellulose using different carbon sources and culture media. *Carbohydrate Polymers*, *117*, p. 518-523.

[166] Cakar, F., et al., (2014). Improvement production of bacterial cellulose by semi-continuous process in molasses medium. *Carbohydrate Polymers*, *106*, p. 7-13.

[167] Cakar, F., et al., (2014). Newly developed medium and strategy for bacterial cellulose production. *Biochemical Engineering Journal*, *92*, p. 35-40.

[168] Keshk, S. M. A. S. & Vitamin, C. (2014). Enhances bacterial cellulose production in Gluconacetobacter. *Carbohydrate Polymers*, *99*, p. 98-100.

[169] Esa, F., Tasirin, S. T. & Rahman, N. A. (2014). Overview of bacterial cellulose production and application. *Agriculture and Agricultural Science Procedia*, *2*, p. 113-119.

[170] Wu, J. M. & Liu, R. H. (2012). Thin stillage supplementation greatly enhances bacterial cellulose production by Gluconacetobacter xylinus. *Carbohydrate Polymers*, *90*, p. 116-121.

[171] Hong, F., et al., (2012). Bacterial cellulose production from cotton-based waste textiles, Enzymatic saccharification enhanced by ionic liquid pretreatment. *Bioresource Technology*, *104*, p. 503-508.

[172] Girones, J., et al., (2007). Blocked isocyanates as coupling agents for cellulose-based composites. *Carbohydrate Polymers*, *68*, p. 537-543.

[173] Koronis, G., Silva, A. & Fontul, M. (2013). Green composites: A review of adequate materials for automotive applications. *Composites Part B: Engineering*, *44*, p. 120-127.

[174] Fonseca-Valero, C., et al., (2015). Mechanical recycling and composition effects on the properties and structure of hardwood cellulose-reinforced high density polyethylene eco-composites. *Composites Part A: Applied Science and Manufacturing*, *69*, p. 94-104.

[175] Rojo, E., et al., (2015). Effect of fiber loading on the properties of treated cellulose fiber-reinforced phenolic composites. *Composites Part B: Engineering*, *68*, p. 185-192.

[176] Thakur, V. K. & Thakur, M. K. (2014). Processing and characterization of natural cellulose fibers/thermoset polymer composites. *Carbohydrate Polymers*, *109*, p. 102-117.

[177] Jawaid, M. & Abdul Khalil, H. P. S. (2011). Cellulosic/synthetic fibre reinforced polymer hybrid composites: A review. *Carbohydrate Polymers*, *86*, p. 1-18.

[178] Akil, H. M., et al., (2011). Kenaf fiber reinforced composites: A review. *Materials & Design*, *32*, p. 4107-4121.

[179] Okereke, M. I., Akpoyomare, A. I. & Bingley, M. S. (2014). Virtual testing of advanced composites, cellular materials and biomaterials: *A review. Composites Part B: Engineering*, *60*, p. 637-662.

[180] Matthews, F. L. & Rawlings, R. D. (1999). *Composite materials: engineering and science.*, Cambridge: Woodhead Publishing Limited.

[181] Dandekar, C. R. & Shin, Y. C. (2012). Modeling of machining of composite materials: A review. *International Journal of Machine Tools and Manufacture*, *57*, p. 102-121.
[182] Varanasi, S., Low, Z. X. & Batchelor, W. (2015). Cellulose nanofibre composite membranes- Biodegradable and recyclable UF membranes. *Chemical Engineering Journal*, *265*, p. 138-146.
[183] Guzman-Puyol, S., et al., (2015). Effect of trifluoroacetic acid on the properties of polyvinyl alcohol and polyvinyl alcohol-cellulose composites. *Chemical Engineering Journal*, *277*, p. 242-251.
[184] Qi, H., et al., (2015). Carbon nanotube-cellulose composite aerogels for vapour sensing. *Sensors and Actuators B, Chemical*, *213*, p. 20-26.
[185] Xu, M., et al., (2015). Highly tough cellulose/graphene composite hydrogels prepared from ionic liquids. *Industrial Crops and Products*, *70*, p. 56-63.
[186] Oun, A. A. & Rhim, J. W. (2015). Preparation and characterization of sodium carboxymethyl cellulose/cotton linter cellulose nanofibril composite films. *Carbohydrate Polymer*s, *127*, p. 101-109.
[187] Oishi, Y., et al., (2015). Structural and mechanical properties of cellulose composites made of isolated cellulose nanofibers and poly(vinylalcohol). *Composites Part A, Applied Science and Manufacturing*, *73*, p. 72-79.
[188] Ifuku, S. & Yano, H. (2015). Effect of a silane coupling agent on the mechanical properties of a microfibrillated cellulose composite. *International Journal of Biological Macromolecules*, *74*, p. 428-432.
[189] Gharbi, A., Hassen, R. B. & Boufi, S. (2014). Composite materials from unsaturated polyester resin and olive nuts residue: The effect of silane treatment. *Industrial Crops and Products*, *62*, p. 491-498.
[190] Zhou, F., Cheng, G. & Jiang, B. (2014). Effect of silane treatment on microstructure of sisal fibers. *Applied Surface Science*, *292*, p. 806-812.
[191] Asumani, O. M. L., Reid, R. G. & Paskaramoorthy, R. (2012). The effects of alkali-silane treatment on the tensile and flexural properties of short fibre non-woven kenaf reinforced polypropylene composites. *Composites Part A: Applied Science and Manufacturing*, *43*, p. 1431-1440.
[192] Wang, J., et al., (2015). Preparation, characterization and application of a novel silane-bridge polyaniline/cotton fiber composite. *Synthetic Metals*, *204*, p. 10-16.
[193] Lu, T., et al., (2013). Effect of surface modification of bamboo cellulose fibers on mechanical properties of cellulose/epoxy composites. *Composites Part B: Engineering*, *51*, p. 28-34.
[194] Siroky, J., et al., (2011). Alkali treatment of cellulose II fibres and effect on dye sorption. *Carbohydrate Polymer*s, *84*, p. 299-307.
[195] Peng, X., et al., (2010). Laccase and alkali treatments of cellulose fibre: Surface lignin and its influences on fibre surface properties and interfacial behaviour of sisal fibre/phenolic resin composites. *Composites Part A: Applied Science and Manufacturing*, *41*, p. 1848-1856.
[196] Foldvary, C. M., Takacs, E. & Wojnarovits, L. (2003). Effect of high-energy radiation and alkali treatment on the properties of cellulose. *Radiation Physics and Chemistry*, *67*, p. 505-508.
[197] Takacs, E., et al., (2000). Effect of combined gamma-irradiation and alkali treatment on cotton-cellulose. *Radiation Physics and Chemistry*, *57*, p. 399-403.

[198] Zhong, J. F., Chai, X. S. & Fu, S. Y. (2012). Homogeneous grafting poly(methyl methacrylate) on cellulose by atom transfer radical polymerization. *Carbohydrate Polymers*, *87*, p. 1869-1873.

[199] Lin, C. X., et al., (2009). Preparation of cellulose graft poly(methyl methacrylate) copolymers by atom transfer radical polymerization in an ionic liquid. *Carbohydrate Polymers*, *78*, p. 432-438.

[200] Hong, K. H., Liu, N. & Sun, H. (2001). UV-induced graft polymerization of acrylamide on cellulose by using immobilized benzophenone as a photo-initiator. *European Polymer Journal*, *45*, p. 2443-2449.

[201] Badawy, S. M., Dessouki, A. M. & Nizam El-Din, H. M. (2001). Direct pyrolysis mass spectrometry of acrylonitrile-cellulose graft copolymer prepared by radiation-induced graft polymerization in presence of styrene as homopolymer suppressor. *Radiation Physics and Chemistry*, *61*, p. 143-148.

[202] Nishino, T., Matsuda, I. & Hirao, K. (2004). All-cellulose composite. *Macromolecules*, *37*(20), p. 7683-7687.

[203] Suryanegara, L., Norio Nakagaito, A. & Yano, H. (2009). The effect of crystallization of PLA on the thermal and mechanical properties of microfibrillated cellulose-reinforced PLA composites. *Composites Science and Technology*, *69*, p. 1187-1192.

[204] Awal, A., Rana, M. & Sain, M. (2015). Thermorheological and mechanical properties of cellulose reinforced PLA bio-composites. *Mechanics of Materials*, *80*, p. 87-95.

[205] Song, Z., Xiao, H. & Zhao, Y. (2014). Hydrophobic-modified nano-cellulose fiber/PLA biodegradable composites for lowering water vapor transmission rate (WVTR) of paper. *Carbohydrate Polymers*, *111*, p. 442-448.

[206] Raquez, J. M., et al., (2013). Polylactide (PLA)-based nanocomposites. *Progress in Polymer Science*, Article In Press.

[207] Jorda-Vilaplana, A., et al., (2014). Surface modification of polylactic acid (PLA) by air atmospheric plasma treatment. *European Polymer Journal*, *58*, p. 23-33.

[208] Lagaron, J. M. (2011). Polylactic acid (PLA) nanocomposites for food packaging applications. *Multifunctional and Nanoreinforced Polymers for Food Packaging*, p. 485-497.

[209] Reis, K. C., et al., (2008). Characterization of polyhydroxybutyrate-hydroxyvalerate (PHB-HV)/maize starch blend films. *Journal of Food Engineering*, *89*, p. 361-369.

[210] Chien, C. C., et al., (2007). Production of poly-b-hydroxybutyrate (PHB) by Vibrio spp. isolated from marine environment. *Journal of Biotechnology*, *132*, p. 259-263.

[211] Patnaik, P. R. (2006). Dispersion optimization to enhance PHB production in fed-batch cultures of Ralstonia eutropha. *Bioresource Technology*, *97*, p. 1994-2001.

[212] Rueda, L., et al., (2013). Cellulose nanocrystal/polyurethane nanocomposites. Study from the viewpoint of microphase separated structure. *Carbohydrate Polymers*, *92*, p. 751-757.

[213] Zavastin, D., et al., (2010). Preparation, characterization and applicability of cellulose acetate-polyurethane blend membrane in separation techniques. *Colloids and Surfaces A: Physicochemical and Engineering Aspects*, *370*, p. 120-128.

[214] Benhamou, K., et al., (2015). Bio-based polyurethane reinforced with cellulose nanofibers: A comprehensive investigation on the effect of interface. *Carbohydrate Polymers*, *122*, p. 202-211.

[215] Sivakumar, M., et al., (1999). Ultrafiltration application of cellulose acetate-polyurethane blend membranes. *European Polymer Journal*, *35*, p. 1647-1651.

[216] Kawahara, J., et al., (2013). Reconfigurable sticker label electronics manufactured from nanofibrillatd cellulose-based self-adhesive organic electronic materials. *Organic Electronics*, *14*, p. 3061-3069.

[217] Valentini, L., et al., (2013). A novel method to prepare conductive nanocrystalline cellulose/graphene oxide composite films. *Materials Letters*, *105*, p. 4-7.

[218] Ummartyotin, S. & Manuspiya, H. (2015). An overview of feasibilities and challenge of conductive cellulose for rechargeable lithium based battery. *Renewable and Sustainable Energy Reviews*, *50*, p. 204-213.

[219] Patel, M. U. M., et al., (2014). Low surface area graphene/cellulose composite as a host matrix for lithium sulphur batteries. *Journal of Power Sources*, *254*, p. 55-61.

[220] Deng, J., et al., (2009). Developments and new applications of UV-induced surface graft *polymer*izations. *Progress in Polymer Science*, *34*, p. 156-193.

[221] Lemus, J., Eguizabal, A. & Pina, M. P. (2015). UV polymerization of room temperature ionic liquids for high temperature PEMs: Study of ionic moieties and crosslinking effects. *International Journal of Hydrogen Energy*, *40*, p. 5416-5424.

[222] Gao, X., et al., (2013). Surface-modified PSf UF membrane by UV-assisted graft polymerization of capsaicin derivative moiety for fouling and bacterial resistance. *Journal of Membrane Science*, *445*, p. 146-155.

[223] Sun, P., et al., (2015). Well-defined cyclopropenone-masked dibenzocyclooctyne functionalized polymers from atom transfer radical polymerization. *Polymer*, *64*, p. 202-209.

[224] Golaz, B., et al., (2012). UV intensity, temperature and dark-curing effects in cationic photo-polymerization of a cycloaliphatic epoxy resin. *Polymer*, *53*, p. 2038-2048.

[225] Gruber, H. F. (1992). Photoinitiators for free radical polymerization. *Progress in Polymer Science*, *17*, p. 953-1044.

[226] Chen, H., et al., (2012). Photopolymerization kinetics of a-dislfone cationic photoinitiator. *Journal of Photochemistry and Photobiology A: Chemistry*, *232*, p. 57-63.

[227] Visconti, M. & Cattaneo, M. (2000). A highly efficient photoinitiator for water-bone UV curable systems. *Progress in Organic Coatings*, *40*, p. 243-251.

[228] Dietliker, K., et al., (2007). Advancements in photoinitiators- Opening up new applications for radiation curing. *Progress in Organic Coatings*, *58*, p. 146-157.

[229] Yano, H., et al., (2005). Optically transparent composites reinforced with networks of bacterial nanofibers. *Advanced Materials*, *17*, p. 153-155.

[230] Nogi, M. & Yano, H. (2008). Transparent nanocomposites based on cellulose produced by bacteria offer potential innovation in the electronics device industry. *Advanced Materials*, *20*, p. 1849-1852.

[231] Juntaro, J., et al., (2012). Bacterial cellulose reinforced polyurethane-based resin nanocomposites: A study of how ethanol and processing pressure affect physical, mechanical and dielectric properties. *Carbohydrate Polymer*s, *87*, p. 2464-2469.

[232] Luthra, S., et al., (2014). Adoption of smart grid technologies: An analysis of interactions among barriers. *Renewable and Sustainable Energy Reviews*, *33*, p. 554-565.

[233] Ho, J. C., et al., (2014). Technological barriers and research trends in fuel cell technologies: A citation network analysis. *Technological Forecasting and Social Change*, *82*, p. 66-79.

[234] Forster, J., et al., (2005). A PVD based barrier technology for the 45 nm node. *Microelectronic Engineering*, *82*, p. 594-599.

[235] Ummartyotin, S., et al., (2011). Si-O barrier technology for bacterial cellulose nanocomposite flexible displays. *Carbohydrate Polymer*s, *86*, p. 337-342.

[236] Genc, S. & Derin, B. (2014). Synthesis and rheology of ferrofluids: a review. *Current Opinion in Chemical Engineering*, *3*, p. 118-124.

[237] Selimefendigil, F., Oztop, H. F. & Al-Salem, K. (2014). Natural convection of ferrofluids in partially heated square enclosures. *Journal of Magnetism and Magnetic Materials*, *372*, p. 122-133.

[238] Aminfar, H., Mohammadpourfard, M. & Maroofiazar, R. (2014). Experimental study on the effect of magnetic field on critical heat flux of ferrofluid flow boiling in a vertical annulus. *Experimental Thermal and Fluid Science*, *58*, p. 156-169.

[239] Poesio, P. & Wang, E. N. (2014). Resonance induced wetting state transition of a ferrofluid droplet on superhydrophobic surfaces. *Experimental Thermal and Fluid Science*, *57*, p. 353-357.

[240] Goharkhah, M. & Ashjaee, M. (2014). Effect of an alternating nonuniform magnetic field on ferrofluid flow and heat transfer in a channel. *Journal of Magnetism and Magnetic Materials*, *362*, p. 80-89.

[241] Ummartyotin, S., et al., (2012). The role of ferrofluid on surface smoothness of bacterial cellulose nanocomposite flexible display. *Chemical Engineering Journal*, 193-194, p. 16-20.

[242] Yang, C., et al., (2009). Piezoelectricity of wet drawn cellulose electro-active paper. *Sensors and Actuators A, Physical*, *154*, p. 117-122.

[243] Kim, J., et al., (2008). Modeling elastic, viscous and creep characteristics of cellulose Electro-Active Paper. *Mechanics of Materials*, *40*, p. 1001-1011.

[244] Kim, H. S., Li, Y. & Kim, J. (2008). Electro-mechanical behavior and direct piezoelectricity of cellulose electro-active paper. *Sensors and Actuators A: Physical*, *147*, p. 304-309.

[245] Yun, S. & Kim, J. (2008). Characteristics and performance of functionalized MWCNT blended cellulose electro-active paper actuator. *Synthetic Metals*, *158*, p. 521-526.

[246] Mahadeva, S., Lee, S. W. & Kim, J. (2008). Effect of heat treatment on the structure, piezoelectricity and actuation behavior of a cellulose electroactive-paper actuator. *Acta Materialia*, *56*, p. 1868-1875.

[247] Choi, H., et al., (2015). Alginate fouling reduction of functionalized carbon nanotube blended cellulose acetate membrane in forward osmosis. *Chemophere*, *136*, p. 204-210.

[248] Badawi, N., et al., (2014). Novel carbon nanotube-cellulose acetate nanocomposite membranes for water filtration applications. *Desilination*, *344*, p. 79-85.

[249] O-Rak, K., et al., (2013). Covalently grafted carbon nanotube on bacterial cellulose composite for flexible touch screen application. *Materials Letters*, *107*, p. 247-250.

[250] Cai, Z., Hou, C. & Yang, G. (2012). Characteristics and bending performance of electroactive polymer blend made with cellulose and poly(3-hydroxybutyrate). *Carbohydrate Polymer*s, *87*, p. 650-657.

[251] Ma, X., Yu, J. & Wang, N. (2008). Glycerol plasticized-starch/multiwall carbon nanotube composites for electroactive polymers. *Composites Science and Technology*, *68*, p. 268-273.

[252] Hong, C. H., et al., (2013). Electroactive bio-composite actuators based on cellulose acetate nanofibers with specially chopped polyaniline nanoparticles through electrospinning. *Composites Science and Technology*, *87*, p. 135-141.

[253] Je, C. H. & Kim, K. J. (2004). Cellophane as a biodegradable electroactive polymer actuator. *Sensors and Actuators A: Physical*, *112*, p. 107-115.

[254] O-Rak, K., et al., (2014). Development of bacterial cellulose and poly(vinylidene fluoride) binary blend system: Structure and properties. *Chemical Engineering Journal*, *237*, p. 396-402.

[255] Lee, J. H., et al., (2015). Grain-size effect on the electrical properties of nanocrystalline indium tin oxide thin films. *Materials Science and Engineering: B*, *199*, p. 37-41.

[256] Skeren, T., et al., (2015). Ion-induced pattern formation on indium tin oxide for alignment of liquid crystals. *Thin Solid Films*, *589*, p. 315-321.

[257] Alvarez-Fraga, L., et al., (2015). Indium-tin oxide thin films deposited at room temperature on glass and PET substrates: Optical and electrical properties variation with the H2-Ar sputtering gas mixture. *Applied Surface Science*, *344*, p. 217-222.

[258] Sivasubramanian, R. & Sangaranarayanan, M. V. (2015). A facile formation of silver dendrites on indium tin oxide surfaces using electrodeposition and amperometric sensing of hydrazine. *Sensors and Actuators B: Chemical*, *213*, p. 92-101.

[259] Al-Oqla, F. M., et al., (2015). Natural fiber reinforced conductive *polymer* composites as functional materials: A review. *Synthetic Metals*, *206*, p. 42-54.

[260] Bagheri, H., Ayazi, Z. & Naderi, M. (2013). Conductive *polymer*-based microextraction methods: A review. *Analytica Chimica Acta*, *767*, p. 1-13.

[261] Al-Saleh, M. H. & Sundararaj, U. (2009). A review of vapor grown carbon nanofiber/*polymer* conductive composites. *Carbon*, *47*, p. 2-22.

[262] Nambiar, S. & Yeow, J. T. W. (2011). Conductive polymer-based sensors for biomedical applications. *Biosensors and Bioelectronics*, *26*, p. 1825-1832.

[263] Ummartyotin, S., et al., (2011). Deposition of PEDOT:PSS nanoparticles as a conductive microlayer anode in OLEDs device by desktop inkjet printer. *Journal of Nanomaterials*, *1*, p. Article ID 606714.

[264] Trajic, J., et al., Raman spectroscopy of ZnS quantum dots. *Journal of Alloys and Compounds*, *637*, p. 401-406.

[265] Shirazi, R., et al., (2014). Origin of low quantum efficiency of photoluminescence of lnP/ZnS nanocrystals. *Journal of Luminescence*, *145*, p. 936-939.

[266] Schmidt, R., Krasselt, C. & Borczyskowski, C. (2012). Change point analysis of matrix dependent photoluminescence intermittency of single CdSe/ZnS quantum dots with intermediate intensity levels. *Chemical Physics*, *406*, p. 9-14.

[267] Ummartyotin, S., et al., (2012). Synthesis and luminescence properties of ZnS and metal (Mn,Cu)-doped-ZnS ceramic powder. *Solid State Sciences*, *14*, p. 299-304.

[268] Ummartyotin, S., et al., (2012). Hybrid organic-inorganic of ZnS embedded PVP nanocomposite film for photoluminescent application. *Comptes Rendus Physique*, *13*, p. 994-1000.

[269] Ummartyotin, S., et al., (2012). Synthesis of colloidal silver nanoparticle for printed electronic. *Comptes Rendus Chimie*, Accepted.

[270] Grande, L., et al., (2012). Graphene for energy harvesting/storage devices and printed electronics. *Particuology*, *10*, p. 1-8.

[271] Huebner, G. & Krebs, M. (2015). *Printed, flexible thin film batteries and other power storage devices*, ed. H.o.F.O. Electronics.

[272] Kumashiro, Y., et al., (2009). Novel materials for electronic device fabrication using ink-jet printing technology. *Applied Surface Science*, *256*, p. 1019-1022.

[273] Ummartyotin, S., Sain, M. & Qvintus, P. (2014). *Cellulose in printed electronic.* Handbook of Green Materials.

INDEX

C

D

E

F

M

N

O

P

Q

R

S

T

U

V

W

X

Y

Z